A MANUAL OF

AMERICAN ENGINEERS & SURVEYORS

INSTRUMENTS

W. & L.E. Gurley's Instrument Manufactory, Established 1845.

W.&L.E. GURLEY,

MANUFACTURERS OF
Civil Engineers'& Surveyors'Instruments.

TROY, N.Y.

A

MANUAL

OF THE PRINCIPAL

INSTRUMENTS

USED IN

AMERICAN ENGINEERING AND SURVEYING.

MANUFACTURED BY

W. & L. E. GURLEY,

TROY, N. Y.

TWENTY-FIRST EDITION.

1874.

THE ASTRAGAL PRESS
Mendham, New Jersey

Library of Congress Catalog Card Number: 92-75615
International Standard Book Number: 1-879335-34-4

Printed in the United States of America

Published by:
THE ASTRAGAL PRESS
5 Cold Hill Road, Suite 12
Mendham, New Jersey 07945-0239

INTRODUCTION

William and Lewis E. Gurley, sons of Ephraim and Clarissa Gurley of Troy, New York, were born in Troy on March 16, 1821, and December 30, 1826. Their father, Ephraim, had been involved in business in the Troy area with now-famous brass founders and instrument makers such as Julius and Truman Hanks, but he died in 1829 and would have had little or no opportunity to teach even his older son William any of the brass or instrument making business. However, by 1839, William, recently graduated from Rensselaer Institute, was making surveying instruments as an apprentice to Oscar Hanks. Five years later, in 1844, his brother Lewis became an apprentice to Jonas H. Phelps and apprenticed to Phelps until he went to college in 1847.

In February, 1845, one year after his brother began his apprenticeship with Jonas Phelps, William and Jonas formed the partnership "Phelps and Gurley". At that time Jonas Phelps had been in the Troy area for 12 years, making mathematical and philosophical (scientific) instruments. In September, 1851, two months after Lewis graduated from Union College in nearby Schenectady, New York, he was made a partner in the Phelps and Gurley firm. Just five months later, in February of 1852, William and Lewis purchased Jonas Phelps' interest in the firm and changed its name to W. & L.E. Gurley; the Gurley name continued for 115 years until the Teledyne Corporation bought the company and continued it as Teledyne Gurley.

The Gurley brothers' company became one of the leading companies in America producing surveying and engineering instruments; it played a major role in the large scale mapping projects that took place in the United States after

1

the Civil War. It is claimed that W. & L.E. Gurley made more instruments during the decade from 1870 to 1880 than in any other decade in their history; I believe that makes the reprint of the 1874 *Gurley Manual* very appropriate.

The first *Gurley Manual* was published in 1855 by William and Lewis E. Gurley. By the time Lewis died in 1897, the W. & L.E. Gurley company had published 32 editions of the *Manual,* including its Semi-Centennial Edition (the 31st) in 1895.

This reprint is of the 21st edition, published and entered in the Office of the Librarian of Congress at Washington in 1874. The edition prior to this one, called the "20th Revised", was issued in 1873 but had been entered (copyrighted) 11 years earlier in 1862 during the Civil War and just before the disastrous fire of May 10, 1862, which destroyed the Gurleys' factory buildings and contents. The Gurleys quickly rebuilt, and the building they moved into in December of 1862 is still owned by Teledyne Gurley today.

The Gurley brothers published at least 15 editions of the 1862 copyright until finally revising the complete manual in 1874; this reprint is of that manual. It should be noted that the cover of the manual read "American Engineers & Surveyors Instruments Twenty First Edition". The title page read "A Manual of the Principal Instruments Used in American Engineering and Surveying"; then in smaller letters continued "Manufactured by W. & L.E. Gurley, Troy, N.Y." The way the cover and title page were structured was no accident and, in my opinion, a stroke of genius by the Gurleys; it led people to believe the manual was about all makes of American surveying and engineering instruments; of course it was not! It also led people to believe it was a manual, but lo and behold, 89 pages of the

2

volume were titled "Supplement to Manual"; the Supplement was a full-fledged catalog advertising hundreds of Gurley products for sale!

When I started surveying in the late 1950's, I was encouraged to believe that every smart surveyor carried a Gurley *Manual* with him in the field because it told how to adjust and calibrate all American-made instruments . . . you can imagine my chagrin and embarrassment when I had it pointed out to me years later that the *Manual* did not have anything in it about any American-made instruments other than Gurleys'.

Nevertheless the Gurley *Manuals* (over 50 editions were printed) have preserved for the student and collector of American scientific instruments an invaluable record of not only what kind of instruments existed but also how they were constructed and used. The *Manuals* also show how, over the 116 years that W. & L.E. Gurley made instruments, the instruments changed, and why. This reprint should be considered a must, not only to scientific instrument students and collectors, but also to "Americana" buffs.

David C. Garcelon
Millbury, Massachusetts
December 1992

PRICE LIST.

Troy, August 15, 1874.

All prices in this work are in U. S. Currency. State what edition of Manual when ordering goods; also give Catalogue Number.

SOLAR COMPASS.

No.   Price.

1.—Solar Compass, with adjusting socket and leveling head tripod.......... $220 00
2.—Micrometer Telescope, 16 to 20 inches long, with rack movement to object glass and movable clips to attach to compass sights..................... 25 00

PATENT SOLAR ATTACHMENT, FOR TRANSITS.

3.—Patent Solar Attachment... $60 00
4.—Vertical Arc. on Silver, with Vernier, movable by tangent screw, reading to 30 Seconds... 20 00

This Solar Attachment can be applied to any Transit, having Level on Telescope clamp and tangent, and the vertical Arc with tangent to Vernier as described above. To parties having the ordinary vertical circle to 1' on Transits made by us, we allow $7 00 for the old circle, and furnish the vertical Arc No. 4 for... 13 00

COMPASSES.

5.—Plain, with Jacob Staff mountings, 4 inch needle, and out-keeper........						$30 00
6.— do	do	do	5	do	"	35 00
7.— do	do	do	6	do	"	40 00
9.—Vernier,	do	do	4	do	"	40 00
10.— do	do	do	5	do	"	45 00
11.— do	do	do	6	do	"	50 00
13.—Railroad,	do	do	single vernier to limb, 5 inch needle......			65 00
14.— do	do	do	do	do	5½	70 00
15.— do	do	do	double	do	5	80 00
16.— do	do	do	do	do	5½	80 00

EXTRAS.

20.—Compass tripod, with cherry legs..................................... $8 00
21.—   do   do   do   Jointed for Mining Engineering 13 00
22.—   do   do   leveling screws and clamp and tangent movements.. 18 00
23.—   do   mountings without legs................................... 7 00
24.—Compound tangent ball.. 6 00
25.—Adjusting socket... 4 00
26.—Jacob Staff mountings, brass head 2 50
27.—   do   do   steel point...................................... 60
28.—Brass cover for compass glass... 1 00

POCKET COMPASSES.

No.		Price.
30.—With folding sights, 2½ inch needle, very serviceable for tracing lines once surveyed		$9 00
31.—With folding sights, 2½ inch needle, with Jacob staff mountings		11 50
32.— " " " 3½ " " " "		13 50
33 — " " " 3½ " " " and two levels		15 00
34.— " " " 3½ " without mountings or levels		11 00
35.—Vernier pocket compass, with staff mountings, two levels, and 3½ inch needle		18 00
36.—Tripod for Pocket Compasses	extra	5 00
Pocket compasses without sights,	See Supplement.	

MINER'S COMPASSES.

MINER'S COMPASS OR DIPPING NEEDLE FOR TRACING IRON ORE.

40.—Glass on both sides, Wood Box, Stop to Needle. See Engraving		$12 00
41.— " " Brass Covers. " "		12 00
42.— " one side, " " " " See Engraving		12 00
(Above Compasses Nos. 40, 41, and 42, without stop to Needle		10 00

MINER'S COMPASS, "NORWEGIAN NEEDLE."

NEW AND BEAUTIFUL ARTICLE.

43.—Miner's Compass, "Norwegian Needle," Glass both sides, Brass Covers, 3 inch Needle. See Engraving $12 00

44.—4 inch "   "   "   "   "   "   "   "   "   "   " 15 00

(Miner's Compasses, No. 43 and No. 44, Norwegian Needle, with stop to Needle, extra 2 00

TRANSITS.

47.—Vernier, plain telescope*, 4 inch needle, with compass tripod							$75 00
48.— do	do	5	do	do	do		80 00
49.— do	do	6	do	do	do		85 C0
50.—Surveyors' do		4 inch single vernier to limb, leveling tripod					135 00
51.— do	do	4½	do	do	do	do	140 00
52.— do	do	5	do	do	do	do	140 00
53.— do	do	5½	do	do	do	do	140 00
54.— do	do	4 inch double vernier to limb,			do		160 00
55.— do	do	4½	do	do	do	do	165 00
56.— do	do	5	do	do	do	do	165 00
57.— do	do	5½	do	do	do	do	165 00
60 —Engineers' do		4 inch double vernier to limb,			do		175 00
61.— do	do	4½	do	do	do	do	180 00
62.— do	do	5	do	do	do	do	180 00
63.— do	do	5 inch, with watch telescope					220 00
64.— do	do	5 inch, with theodolite axis					220 00

* A "plain" telescope is one without any of the attachments or extras, as we term them, such as the clamp and tangent, vertical circle and level.

EXTRAS TO TRANSITS.

No.   Price.

67.—Vertical circle, 3½ inch diameter, vernier reading to five minutes........ $8 00
68.—   "   4½   "   "   "   single " 14 00
68¼.—Vertical Arc, 6 inch diameter divided on silver, with vernier, movable
        by tangent screw, reading to 30 seconds. See No. 4...................20 00
69.—Clamp and tangent movement to axis of telescope........................ 7 00
70.—Level on telescope, with ground bubble and scale................. .. .... 14 00
71.—Rack and pinion movement to eye-glass.................................. 5 00
72.— Sights on telescope, with folding joints............. .... 8 00
73.—Sights on standards at right angles to telescope.......................... 8 00
74.—Jointed Tripod legs, for Mining Engineering.......................extra 5 00

LEVELING INSTRUMENTS.

75.—Sixteen-inch telescope, with leveling tripod........................... $135 00
76.—Eighteen   "   "   " 135 00
77.—Twenty   "   "   " 135 00
78.—Twenty-two   "   "   " 135 00
79.—Fifteen-inch dumpy, or builder's level, with leveling tripod.............. 75 00
80.—Eleven   "   "   "   " 60 00

HAND LEVELS.

85.—Locke's Hand Level, made of German Silver. (See Nos. 1075 and 1076).. $12 00
86.—   do   do   do   brass   do   do   do   ... 10 00

LEVELING RODS, &c.

90.—Philadelphia Rod.. $16 00
91.—Yankee or Boston.. ... 16 00
92.—New York, with improved mountings..................................... 16 00
93.—Mountings for New York rod, target..................................... 5 50
94.—   "   "   " clamp........... .... 2 50
95.—Simple Rod with Target, reading to 1,000ths of a foot, rod either 8 or 10 feet
        long.. 5 00

FLAG STAFFS.

93.—6 feet long, with steel pointed shoe, and divided off in feet, which are
        painted red and white, alternately $3 00
97.—8 feet long, with steel pointed shoe, and divided off in feet, which are
        painted red and white, alternately...................................... 3 25
98.—10 feet long, with steel pointed shoe, and divided off in feet, which are
        painted red and white, alternately..................................... 3 50

POCKET ANEROID BAROMETERS,

for ascertaining heights, differences of level and meteorological changes, approach of storms, &c.

    These instruments as now made are nearly as portable as an ordinary watch, and yet are fully as accurate as the larger sizes. They are of very great service to the engineer and tourist, as well as to the scientific observer, and are rapidly coming into general use.

    The ordinary styles indicate altitudes to 8,000 feet, and can be furnished, reading as high as 18,000 and 20,000 feet.

    They are all inclosed in neat morocco cases, and are accompanied by a hand-book of instructions.

POCKET ANEROID BAROMETERS.—*Continued.*

No.						Price.
100.—Gilt Aneroid Barometer 1¾ in., Enamel Dial. No altitude...............						$16 00
101.—	Do	do	do	Silvered	do	20 00
102.—	Do	do	do	do	do fine.............	30 00
103.—	Do	do	do	do	do Curved Thermometer	22 00
104.—	Do	do	do	do	do do open face	23 00
105.—	Do	do	do	do	8,000 feet....................	23 00
106.—	Do	do	do	do	10,000 ft. Compensate d	28 00
107.—	Do	do	do	do	10,000 ft. do RevolvingRing	24 00
108.—	Do	do	do	do	10,000 ft. Compensated, Curved Thermometer, Revolving Ring	27 00
109.—	Do	do	do	do	10,000 feet. do do do	33 00
110.—	Do	do	1¼	do	8,000 feet, Compensated......	35 00
111.—Silver		do	1¾	do	No altitude, do	37 00
112.—Gilt Do		do	2¾	do	do	19 00
113.—	Do	do	do	do	do Curved Ther. Open.	23 00
114.—	Do	do	do	do	8,000 feet....................	24 00
115.—	Do	do	do	do	10,000 feet, Compensated, Curved Thermometer, Revolving Ring.	26 00
116.—	Do	do	do		20,000 ft. Compensated, Revolving Ring	28 00

CHAINS.

120.—100 feet, with oval rings, No. 5 refined iron wire......................							$12 00
121.—	"	"	"	6	"	"	.... 9 00
122.— 50 feet,	"	"	5	"	"	.... 6 50
123.— 50 feet,	"	"	6	"	"	.... 5 00
124.— 66 feet,	"	"	8	"	"	.... 4 75
125.— 33 feet,	"	"	8	"	"	.... 2 75
126.— 66 feet,	"	"	10	"	"	.... 4 00
127.— 33 feet,	"	"	10	"	"	.... 2 50
130.—100 feet,	"	"	8 best steel wire...........................			12 00
131.—100 feet,	"	"	10	"	"	.... 10 50
132.— 50 feet,	"	"	8	"	"	.... 6 50
133.— 50 feet,	"	"	10	"	"	.... 5 75
134.— 66 feet,	"	"	8	"	"	.... 10 50
135.— 66 feet,	"	"	10	"	"	.... 8 00
136.— 33 feet,	"	"	8	"	"	.... 5 75
137.— 33 feet,	"	"	10	"	"	.... 4 50
138.—100 feet, brazed links and rings, No. 12 best steel wire, tempered........							15 00
139.— 50 feet,	"	"	12	"		.... 8 00
140.— 66 feet,	"	"	12	"		.... 14 00
141.— 33 feet,	"	"	12	"		.... 7 00

SPANISH OR MEXICAN VARA CHAINS.

FOR USE IN TEXAS, MEXICO, SOUTH AMERICA, AND CUBA.

145.—10 varas, 50 links, No. 10 refined iron wire,...........................						$2 50
146.—20 Do.	100 do.	No. 10	do.	do.	....	4 00
147.—10 Do	50 do.	No. 8	do.	do.	....	2 75
148.—20 Do.	100 do.	No. 8	do.	do.	....	4 75

SPANISH OR MEXICAN VARA CHAINS.—*Continued.*

No.						Price.
149.—10 varas, 50 links, No. 10 best steel wire						$4 50
150.—20 Do.	100 do.	No. 10	do.	do.		8 00
151.—10 Do.	50 do.	No. 8	do.	do.		5 75
152.—20 Do.	100 do.	No. 8	do.	do.		10 50
153.—10 Do.	Brazed links and rings No. 12 steel wire, tempered					7 00
154.—20 Do.	do.	do.	No. 12	do.	do.	14 00

The price of Vara chains of 40 and 80 links, is the same as that of 50 and 100 links.

155.—Metre chains made to order at the same price, as corresponding length American chains...

156.—Pennsylvania chains of 2 and 4 poles with 40 and 80 links, same price as chains of 50 and 100 links.....................................

Steel Snaps to make any of above " Half Chains," no extra charge.

GRUMMAN PATENT CHAINS.

Drag Chains.

No.								Price.
160.—66 feet, No. 15 tempered steel wire, 100 links, weight 1¼ lbs,								$10 00
							With 10 extra links.	
161.—33 Do.	15	do.	50	do.	¾			6 00
							With 5 extra links.	
162.—100 Do.	15	do.	200	do.	2 lbs			14 00
							With 15 extra links.	
163.—50 Do.	15	do.	100	do.	1 lb			8 00
							With 10-extra links.	
164.— 33 feet. No, 12 wire, 5 tallies, with 5 extra links, 1⅛ lbs,								7 00
165.— 66 "	"	10 "	10	"	3 "			14 00
166.— 50 "	"	5 "	5	"	2⅛ "			8 00
167.—100 "	"	10 "	10	"	4⅛ "			15 00

168.— 50 feet, No. 18 tempered steel wire, 100 links, with attachments of spring-balance, level, and thermometer, for very accurate measurements ; weight ¾ lb.. $17 00

169.—Set of 10 Marking Pins, very light, with leather case.................... 2 00

170.—Brass Plummet, to use with light chain................................. 2 00

171.—Lead   "   "   "   " 1 50

172.—Spring-balance to use with either of above chains...................... 2 00

MARKING PINS.

No.				Price.
175.—Set of 11 Pins, iron wire, No. 4				$1 50
176.— "	"	steel wire, No. 6		2 00
177.— "	"	brass wire, No. 4		3 00
178.— "	"	steel wire, loaded		3 00

CHESTERMAN'S METALLIC TAPE MEASURES.

These tapes are made of linen thread, interwoven with fine brass wire, not so liable to stretch as the usual linen tape, and better calculated to withstand the effect of moisture. They are in substantial leather cases.

CHESTERMAN'S METALLIC TAPE MEASURES.—*Continued.*

No.							Price.
180.—Metallic Tape Measure, 24 feet long, in 10ths or 12ths, each,							$2 40
181.	Do.	do.	33	do.	do,	do. do.	2 75
182.	Do.	do.	50	do.	do.	do. do.	3 50
183.	Do.	do.	66	do.	do.	do. do.	4 00
184.	Do.	do.	75	do.	do.	do. do.	4 50
185.	Do.	do.	80	do.	do.	do. do.	4 75
186.	Do.	do.	100	do.	do.	do. do.	5 50

CHESTERMAN'S METALLIC TAPES WITHOUT BOXES.

187.—Chesterman's Metallic Tapes, without box, 50 feet, 10ths or 12ths						$2 00
188.	Do.	do.	do.	66	do.	2 25
189.	Do.	do.	do.	75	do.	2 50
190.	Do.	do,	do.	100	do.	3 50

CHESTERMAN'S STEEL TAPE MEASURES.

Steel Tape Measures; all steel, to wind up in a box, same as linen measures, the most accurate, durable and portable measures.

191.—Steel Tape Measure, 10 feet long, in 10ths or 12ths, in German Silver case, each,							$4 00
192.—Steel Tape Measure, 10 feet long, tape divided on one side to 12ths, and on the other to centimetres and milimetres,							4 25
193.—Steel Tape Measure, 25 feet long, in 10th or 12ths, each,							6 00
194.	Do.	do.	33	do.	do.	do. do.	7 50
195.	Do.	do.	40	do.	do.	do. do.	8 50
196.	Do.	do.	50	do.	do.	do. do.	10 00
197.	Do.	do.	66	do.	do.	do. do.	13 00
198.	Do.	do.	75	do.	do.	do. do.	15 00
199.	Do.	do.	100	do.	do.	do. do.	19 00

CHESTERMAN'S STEEL POCKET TAPE MEASURES.

200.—Chesterman's steel pocket tapes, in German silver cases, with spring and stop. Tapes divided in 10ths or 12ths of feet, 3 feet long,						$2 00
201.	do.	do.	do.	4	do.	2 25
202.	do.	do.	do.	5	do.	2 50
203.	do.	do.	do.	6	do.	2 75

These Pocket Tapes, with divisions to centimeters and millimeters on the other side, 25 cents per tape higher.

PAINE'S PATENT STANDARD STEEL TAPES.

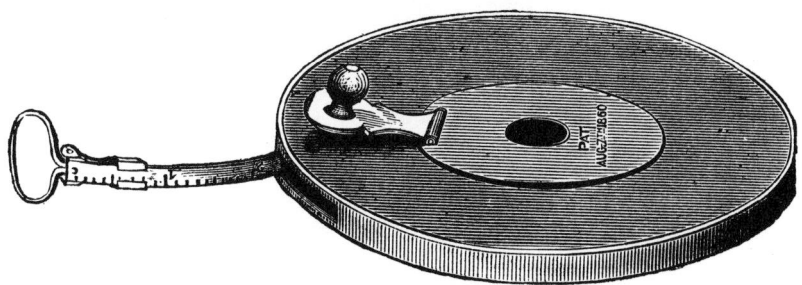

No. 206.

No.							Price.
206.—Standard Steel Tapes, in Japanned case, 25 feet long, 10ths or 12ths,							$3 50
207.	Do.	do.	do.	33	do.	do.	4 50
208.	Do.	do.	do.	50	do.	do.	6 00
209.	Do.	do.	do.	66	do.	do.	8 00
210.	Do.	do.	do.	75	do.	do.	10 00
211.	Do.	do.	do.	100	do.	do.	12 00

PAINE'S PATENT STANDARD STEEL TAPES.

IN LEATHER CASES, FLUSH HANDLES.

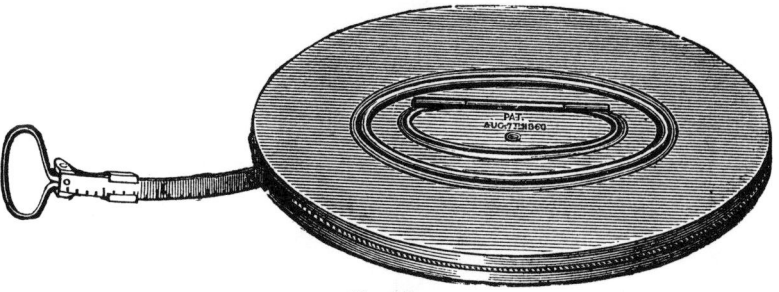

No. 212.

212.—Steel Tape Measure, 33 feet long, 10ths, or 12ths					$5 50
213.	Do.	do.	50	do,	do.	8 00
214.	Do.	do.	66	do.	do.	10 00
215.	Do.	do.	75	do.	do.	12 00
216,	Do.	do.	100	do.	do.	15 00

EXTRAS TO PAINE'S PATENT STANDARD STEEL TAPES.

217.—Handles, with graduated scale, per pair,............................... $3 00
218.—Pocket Thermometers,.. .... 1 50
219.—Spring Balance and Level,...................................... .... 4 00

Information to Purchasers.

INSTRUMENTS WANTED.—In regard to the best kind of instruments for particular purposes, we would here say, that where only common surveying, or the bearing of lines in the surveys for county maps is required, a plain compass is all that is necessary. In cases where the variation of the needle is to be allowed, as in retracing the lines of an old survey, &c., the vernier compass or the vernier transit is required.

Where, in addition to the variation of the needle, horizontal angles are to be taken, and in cases of local attraction, the railroad compass is preferable; and for a mixed practice of surveying and engineering, we consider the surveyor's transit superior to any instrument made by us or any other manufacturers.

In the surveys of U. S. public lands, the county and township lines are required to be run by such instruments as the solar compass.

Where engineering is the exclusive design, the engineer's transit and the leveling instrument are of course indispensable.

The builders' level is intended for laying out mill seats and determining the levels of buildings in course of erection.

WARRANTY.—All our instruments are examined and tested by us in person, and are sent to the purchaser adjusted and ready for immediate use.

They are warranted correct in all their parts—we agreeing in the event of any defect appearing after reasonable use, to repair or replace with a new and perfect instrument, promptly and at our own cost, express charges included, or we will refund the money and the express charges paid by the customer.

Instances may sometimes occur, in a business as large and widely extended as ours, where, owing to careless transportation, or to defects escaping the closest scrutiny of the maker, instruments may reach our customers in bad condition. We consider the retention of such instruments in all cases an injury very much greater to us than to the purchaser himself.

TRIAL OF INSTRUMENTS.—It may often happen that this statement of the prices and quality of our instruments may come into the hands of those who are entirely unacquainted with us, or with the quality of our work, and who therefore feel unwilling to make a final purchase of an article, of the excellence of which they are not perfectly assured.

To such we make the following proposition: We will send the instrument to the express station nearest the person giving the order, and direct the express agent, on delivery of the same, to collect our bill, together with charges of transportation, and hold the money on deposit until the purchaser shall have had, say two weeks, actual trial of its quality

If not found as represented, he may return the instrument before the expiration of that time, and receive the money paid, in full, including express charges, and direct the instrument to be returned to us.

EXTENT OF OUR BUSINESS.—The manufacture of surveying instruments has been conducted by us over thirty years, and thousands of our instruments have been distributed to customers in all parts of the United States and Canadas; in Cuba, South America, Sandwich Islands, and Japan.

Our facilities for manufacturing, which for many years have been far superior to those of any other similar establishment, we have now (1874) greatly increased by the introduction of new machinery and tools of the most improved construction. Our manufactory has been re-built of nearly three times its former size, and we are better prepared than ever before to fill orders for any of our instruments with promptness and satisfaction.

LOW PRICES OF OUR INSTRUMENTS.—It is often urged by other makers. and persons prejudiced in their favor, that it is impossible to make first rate instruments at the prices charged by us, and which are so very far below those of other skillful manufacturers.

We have only to reply, in addition to what we have stated in our warranty, that a visit to our works, and a comparison of our facilities, with those of our competitors, would dispel all questions as to our ability to surpass them, not only in the cheapness, but also in the superior quality of our work.

PACKING, &c.—Each instrument is packed in a well finished mahogany case, furnished with lock and key and brass hooks, and leather strap for convenience in carrying. Each case is provided with screw drivers, adjusting pin, and wrench for centre pin, and, if accompanied by a tripod, with a brass plumb-bob; with all instruments for taking angles without the needle, a reading microscope is also furnished.

Unless the purchaser is already supplied, each instrument is accom
nied by our " Manual," giving full instructions for such adjustments and
repairs as are possible to one not provided with the facilities of an instrument maker.

When sent to the purchaser, the mahogany cases are carefully enclosed
in outside packing boxes, of pine, made a little larger on all sides to
allow the introduction of elastic material, and so effectually are our instruments protected by these precautions, that of several thousand sent out
by us during the last thirty years, in all seasons, by every mode of
transportation, and to all parts of the Union and the Canadas, not more
than three or four have sustained any serious injury.

MEANS OF TRANSPORTATION.—Instruments can be sent by express to
almost every town in the United States and Canadas, regular agents
being located at all the more important points, by whom they are forwarded to smaller places by stage. The charges of transportation from
Troy to the purchaser are in all cases to be borne by him, we guaran-
teeing the safe arrival of our instruments to the extent of express transportation, and holding the express companies responsible to us for all
losses or damages on the way.

FINISH OF INSTRUMENTS.—Customers ordering instruments, will do us a
favor by mentioning whether they prefer them of bright, or bronze finish,
the cost being the same in either case.

If no direction is given, we usually send instruments of bronze finish.

TERMS OF PAYMENT are uniformly cash, and we have but one price,
whether ordered in person or by mail. Our terms are as low as we think
instruments of equal quality can be made, and will not be varied from the
list given on the previous pages.

Remittances may be made by a draft, payable to our order at Troy,
Albany, New York, Boston or Philadelphia, which can be procured from
banks or bankers in almost all the larger villages, or by post office money
order.

These may be sent by mail with the order for the instrument, and if
lost or stolen on the route, can be replaced by a duplicate draft, obtained
as before, and without additional cost.

The customer may also send the money in advance through the express
agent, or as is most common, may pay the agent on receipt of the instrument in funds current in New York or Boston.

The cost of returning the money on bills collected by express of amounts
under $20, will be charged to the customer.

REPAIR OF INSTRUMENTS.

Hundreds of instruments of our own and others' make, come to us every year for refitting and repairs, and so much correspondence arises therefrom, that we are led to believe that a brief statement in this place, of the cost of such repairs, &c., will be of service to our customers and ourselves.

Most instruments sent to us for repairs are injured by falls; many are worn and defective in parts after long use; and others are sent for repolishing and renovation.

We advise our customers having instruments in need of repairs, &c., to send them immediately to us, as our facilities enable us to do the work much more economically and promptly than any other maker however accessible.

They should always, when practicable, be placed in their own boxes, and these inclosed in an outside packing case, an inch larger in all its dimensions, that the interval between the two may be filled with paper wadding, hay or fine shavings.

A note specifying the repairs needed, should accompany the instrument, and a letter should also be sent by mail to us, giving not only directions as to the repairs, but also stating when the return of the instrument is required, and the precise location to which it should be forwarded. It should also be remembered that each instrument is made to fit its own spindle and no other; and therefore this part with the parallel plates and leveling screws, if it has any, should always be sent with it.

The legs and brass head in which they are inserted need never be sent, unless themselves in need of repairs.

COMPASSES.—These come to us with the plates sprung, the sights bent or broken, the glass or level vials fractured, and the pivot so dulled as to render the needle sluggish and unreliable. The cost of repairing the defects above named, ranges from 2 to 8 or 10 dollars. A new pair of sights fitted costs 5 dollars; a new needle, with jeweled centre and pivot complete, $2.50; a new jeweled centre, $1.50.

The compass should always be accompanied by the ball spindle, and if a new ball spindle is required, the whole instrument, or at least the socket in which the spindle fits, should be sent with the letter of advice to us; a new ball spindle costs two dollars.

TRANSIT INSTRUMENTS.—The repairs of the Vernier Transits cost about the same as those of the compasses above stated.

The injuries sustained by the falls of Engineers' and Surveyors' Transits are usually much more serious; in these the plates, standards and cross-bar of telescope are often bent, and the sockets or centres usually so de-ranged as to be entirely useless.

The cost of repairing an instrument with such injuries, ranges from 10 to 30 or even 50 dollars, the new sockets alone costing from 15 to 20 dollars.

LEVELING INSTRUMENTS are generally much less injured by falling than Transits, the damages being included usually in the bending of the cross-bar, the springing of the sockets, and the breaking of the level vial.

The cost of repairs varies from 5 to 15 dollars; a new level vial set in the tube costs two dollars.

RE-POLISHING INSTRUMENTS.—The cost of re-polishing an instrument, involving also of course its complete renovation and adjustment, varies with the different kinds, but may be stated generally as follows:

Compasses, from......................... .. $5 to $8
Transits   do .... 10 to 16
Levels   do .... 10 to 13

No additional charge is made for bronzing or blackening an instrument when re-polished.

PAYMENT OF REPAIRS, &c., may be made at the express office where the instrument is received, the customer paying for the first transportation of the instruments to us or not as he may prefer. Whenever the freight is paid in advance, the express receipt should be mailed immediately to us.

W. & L. E. GURLEY,
Mathematical Instrument Makers,
FULTON ST., OPPOSITE NORTH END OF UNION R. R. DEPOT, TROY, N. Y.

PREFACE

TO THE TWENTY-FIRST EDITION.

———•∞∞•———

WE herewith present the Engineers and Surveyors of the Union, this new edition of our little work, materially enlarged, and, as we trust, improved.

We are now much better furnished with facilities of all kinds to prosecute with enlarged success the business which we have conducted over thirty years.

It is with the hope, therefore, that we shall still further enlarge the list of our many patrons and friends in this and in other countries, and that this description of our Instruments may be of increasing service to the Surveyor and Engineer, that we now commit it to their indulgence.

W. & L. E. GURLEY.

TROY, August 15th, 1874.

Surveying Instruments.

THE various instruments used in Surveying may be conveniently arranged, into two general divisions.

(1.) NEEDLE instruments,—or such as owe their accuracy and value to the magnetic needle only, embracing the Plain and Vernier compasses, and the Vernier Transit.

(2.) ANGULAR instruments, including those in which the horizontal angles, are measured by a divided circle and verniers, as well as by the needle also; as the Railroad Compass, the Surveyors' and Engineers' Transits, &c.

In the present work we shall consider first, those instruments comprised in the first division, and, as in these the accuracy of the horizontal angles indicated, depends upon the delicacy of the needle, and the constancy with which it assumes a certain direction, termed the "magnetic meridian," we shall here remark briefly upon the *form*, the *length*, and the *movement* of

THE MAGNETIC NEEDLE.—The *forms* of the needle are almost infinitely varied, according to the taste or fancy of the maker or surveyor, but may be resolved into two general classes, one having the greatest breadth in a horizontal, the other in a vertical direction.

We have usually made our needles about one-twentieth of an inch broad and one-third as thick, parallel from end to end, the north and south poles being distinguished from each other, by a small scollop on the north end.

Of course the form of the needle is always varied according to the choice of our customers, and without additional charge.

The *length* of the needle varies in different instruments, from four to six or even seven inches, those of five and a half, or six inches long, being generally preferred by surveyors.

The *movement* of the needle, with the least possible friction, is secured by suspending it, by a steel or jewel centre, upon a hardened stee¹ pivot, the point of which is made perfectly sharp and smooth.

The test of the delicacy of a magnetic needle is the number of horizontal vibrations, which it will make in a certain arc, before coming to rest—besides this most surveyors prefer also to see a sort of quivering motion in a vertical direction

This quality, which is manifested more in a horizontal, than in a vertical needle, and depends upon the near coincidence of the point of suspension with the centre of gravity of the needle, serves to show merely that the cap below is unobstructed.

Having now considered the different qualities of a good needle, we shall proceed to speak of those instruments of which it makes so important a part; of these, the most simple is that termed the

PLAIN COMPASS.

Fig. 1.

As represented above, the Plain Compass has a needle six

inches long, a graduated circle, main plate, levels and sights, and is placed upon the brass head of the "Jacob staff."

THE COMPASS CIRCLE in this, as in all our instruments, is divided to half degrees on its upper surface, the whole degree marks being also cut down on the inside circumference, and is figured from 0 to 90, on each side of the centre or "line of zeros."

The circle and face of the compass are silvered.

THE SPIRIT LEVELS are placed at right angles to each other so as to level the plate in all directions, and are balanced upon a pivot underneath the middle of the tube, so as to be adjustable by a common screw-driver.

THE SIGHTS, or standards, have fine slits cut through nearly their whole length, terminated at intervals by large circular apertures, through which the object sighted upon is more readily found. Sometimes a fine horse-hair or wire is substituted for one half the slit, and placed alternately with it on opposite sights.

TANGENT SCALE.—The right and left hand edges of the sights of our compasses, have respectively an eye-piece, and a series of divisions, by which angles of elevation and depression, for a range of about twenty degrees each way, can be taken with considerable accuracy.

Such an arrangement is very properly termed a "tangent scale," the divided edges of the north sight, being tangents to segments of circles having their centres at the eye-pieces, and their points of contact with the tangent lines at the zero divisions of the scale.

The cut shows the eye-piece and divisions for angles of depression; those for angles of elevation, concealed in this cut, are seen in that of the Railroad Compass.

THE JACOB STAFF mountings which are furnished with all our compasses, and packed in the same case, consist of the brass head already mentioned, and an iron ferule or shoe, pointed with steel, so as to be set firmly in the ground.

The staff, to which the mountings should be securely fastened, is procured from any wheelwright, or selected by the surveyor himself from a sapling of the forest.

To adjust the Compass.

THE LEVELS.—First bring the bubbles into the centre, by the pressure of the hand on different.parts of the plate, and then turn the compass half way around ; should the bubbles run to the end of the tubes, it would indicate that those ends were the highest ; lower them by tightening the screws immediately under, and loosening those under the lowest ends until, by estimation, the error is half removed ; level the plate again, and repeat the first operation until the bubbles will remain in the center, during an entire revolution of the compass.

THE SIGHTS may next be tested by observing through the slits a fine hair or thread, made exactly vertical by a plumb. Should the hair appear on one side of the slit, the sight must be adjusted by filing off its under surface on that side which seems the highest.

THE NEEDLE is adjusted in the following manner : Having the eye nearly in the same plane with the graduated rim of the compass circle, with a small splinter of wood or a slender iron wire, bring one end of the needle in line with any prominent division of the circle, as the zero, or ninety degree mark, and notice if the other end corresponds with the degree on the opposite side ; if it does, the needle is said to "cut" opposite degrees ; if not, bend the centre-pin by applying a small brass wrench, furnished with our compasses, about one eighth of an inch below the point of the pin, until the ends of the needle are brought into line with the opposite degrees.

Then holding the needle in the same position, turn the compass half way around, and note whether it now cuts opposite degrees ; if not, correct half the error by bending the needle, and the remainder by bending the centre-pin.

The operation should be repeated until perfect reversion is secured in the first position.

This being obtained, it may be tried on another quarter of the circle ; if any error is there manifested, the correction must be made in the centre-pin only, the needle being already straightened by the previous operation.

When again made to cut, it should be tried on the other quarters of the circle, and corrections made in the same manner until the error is entirely removed, and the needle wil' reverse in every point of the divided surface

To use the Compass.

In using the compass the surveyor should keep the south end towards his person, and read the bearings from the north end of the needle. He will observe that the E and W letters on the face of the compass are reversed from their natural position, in order that the direction of the line of sight may be correctly read.

The compass circle being graduated to half degrees, a little practice will enable the surveyor to read the bearings to quarters, or even finer—estimating with his eye the space bisected by the point of the needle, and as this is as low as the traverse table is usually calculated, it is the general practice.

Sometimes, however, a small vernier is placed upon the south end of the needle, and reads the circle to five minutes of a degree—the circle being in that case graduated to whole degrees.

This contrivance, however, is quite objectionable on account of the additional weight imposed on the centre-pin and the difficulty of reading a vernier which is in constant vibration, and is therefore but little used.

To TAKE ANGLES OF ELEVATION.—Having first leveled the compass, bring the south end towards you, and place the eye at the little button, or eye piece, on the right side cf the south sight, and with the hand fix a card on the front surface of the north sight, so that its top edge will be at right angles to the divided edge, and coincide with the zero mark ; then sighting over the top of the card, note upon a flagstaff the height cut by the line of sight ; then move the staff up the elevation, and carry the card along the sight until the line of sight again cuts the same height on the staff, read off the degrees and half degrees passed over by the card, and we shall have the angle required.

FOR ANGLES OF DEPRESSION.—Proceed in the same manner, using the eye-piece and divisions on the opposite sides of the sights, and reading from the top of the standards.

JACOB STAFF SOCKET.—The compass is furnished with a ball spindle, or socket, upon which it turns, and by which it is levelled. The ball may be placed in a single or "jacob staff" socket, as represented in the figure, or in a compass tripod, such as is shown in the cut of the Vernier Transit beyond.

CLAMP SCREW.—In the side of the hollow cylinder, or socket of the compass, which fits to the ball spindle, is a screw by which the instrument may be clamped to the spindle in any position.

SPRING CATCH.—Besides the clamp screw, we now have fitted to the sockets of our compasses a little spring catch which, as soon as the instrument is set upon the spindle, slips into a groove, and thus removes all danger of falling when the instrument is carried.

NEEDLE LIFTER.—There is also underneath the main plate a needle lifting screw which, by moving a concealed spring raises the needle from the pivot, and thus prevents the blunting of the point in transportation.

When the compass is not in use it is the practice of many surveyors to let down the needle upon the point of the centre-pin, and let it assume its position in the magnetic meridian, so as to retain or even increase its polarity.

We would advise in addition, that after the needle has settled it should be raised against the glass, in order not to dull the point of suspension.

OUTKEEPER.—A small dial plate, having an index turned by a milled head underneath, is often used with this and the other compasses to keep tally in chaining.

The dial is figured from 0 to 16, the index being moved one notch for every chain run.

ELECTRICITY.—A little caution is necessary in handling the compass, that the glass covering be not excited by the friction of cloth, silk, or the hand, so as to attract the needle to its under surface.

A brass cover is sometimes fitted over the glass of the compass, and serves to protect it from accident, as well as to prevent electric disturbance.

When, however, the glass becomes electric, the fluid may be removed by breathing upon it, or touching different parts of its surface with the moistened finger.

An ignorance of this apparently trifling matter has caused many errors and perplexities in the practice of the inexperienced surveyor.

Repairs of the Compass.

To enable the surveyor to make such repairs as are possible without having recourse to an instrument maker, we here add a few simple directions.

1. THE NEEDLE.—It may sometimes happen that the needle

has lost its polarity, and needs to be re-magnetized; this is effected in the following manner:

The operator being provided with an ordinary permanent magnet,* and holding it before him, should pass with a gentle pressure each end of the needle from centre to extremity over the magnetic pole, describing before each pass a circle of about six inches radius, to which the surface of the pole is tangent, drawing the needle towards him, and taking care that the north and the south ends are applied to the *opposite* poles of the magnet.

Should the needle be returned in a path near the magnetic pole, the current induced by the contact of the needle and magnet, in the pass just described, would be reversed, and thus the magnetic virtue almost entirely neutralized at each operation.

When the needle has been passed about twenty-five times in succession, in the manner just described, it may be considered as fully charged.

A fine brass wire is wound in two or three coils on the south end of the needle, and may be moved back or forth in order to counterpoise the varying weight of the north end.

2. THE CENTRE PIN.—This should occasionally be examined, and if much dulled, taken out with the brass wrench, already spoken of, or with a pair of plyers, and sharpened on a hard oil stone—the operator placing it in the end of a small stem of wood, or a pin vice, and delicately twirling it with the fingers as he moves it back and forth at an angle of about 3C deg. to the surface of the stone.

When the point is thus made so fine and sharp as to be invisible to the eye, it should be smoothed by rubbing it on the surface of a soft and clean piece of leather.

3. To PUT IN A NEW GLASS.—Unscrew the "bezzle ring" which holds it, and with the point of a knife blade spring

---

* A magnet suitable for this purpose costs 25 to 50 cents.

out the little brass ring above the glass. remove the old glass and scrape out the putty ; then if the new glass does not fit, smooth off its edges by holding it obliquely on the surface of a grind stone until it will enter the ring easily ; then put in new putty, spring in the brass ring, and the operation will be complete.

4. To REPLACE A SPIRIT LEVEL.—Take out the screws which hold it on the plate, pull off the brass ends of the tube, and with a knife blade scrape out the plaster from the tube; then with a stick made a little smaller than the diameter of the tube, and with its end hollowed out, so that it will bear only on the broad surface of the level vial, push out the old vial and replace it with a new one, taking care that the crowning side, which is usually marked with a file on the end of the vial, is placed on the upper side.

When the vial does not fit the tube it must be wedged up by putting under little slips of paper until it moves in snugly.

After the vial is in its place, put around its ends a little boiled plaster, mixed with water to the consistency of putty, taking care not to allow any to cover the little tip of the glass, then slip in the brass ends and the operation will be completed.

A little beeswax, melted and dropped upon the ends of the vial, is equally as good as the boiled plaster, and often more easily obtained.

We would here remark that an extra glass and level vials are always furnished, free of charge, with our instruments, whenever desired by the purchaser.

Sizes of the Plain Compass.

Three different sizes of this instrument are in common use, having respectively four, five and six-inch needles, and dif- fering also in the length cf the main plate, which in the foui inch compass is twelve and a half inches long, and in the larger sizes, fifteen and a half inches.

The six-inch needle compass is generally preferred.

Weight of the Plain Compasses.

The average weights of the different sizes, with the brass mountings of the jacob staff, are :

> For the 4-inch needle, 5½ lbs.
> For the 5-inch needle, 6½ lbs.
> For the 6-inch needle, 8 lbs.

The plain compass, which was the only one in use in this country previous to the time of David Rittenhouse, has gradually given way to the superior advantages of the Vernier or Rittenhouse compass, which we shall now proceed to describe

VERNIER COMPASS.

6 INCH

Made by

W. & L. E. GURLEY,

TROY, N.Y.

Price as shown above $50.00.

Surveying Instruments.

THE VERNIER COMPASS.

This instrument, represented in the engraving opposite, is in most respects like that already described, differing from it mainly in having its compass circle, to which is attached a " vernier," movable about a common centre a short distance in either direction, thus enabling the surveyor to set the zeros of the circle at any required angle with the line of sights, the number of degrees contained in this angle or the " variation of the needle " being read off by the vernier.

The movement of the circle is effected either by a slow moving or " tangent screw," as shown in the engraving, or by a concealed rack and pinion—the head of which projects from the under side of the main compass plate.

When the variation is set off as described, the circle is securely fastened in its position by a clamping nut underneath the main plate.

BALL SPINDLE.—The compass is usually fitted to a spindle made slightly conical and having on its lower end a ball turned perfectly spherical, and confined in a socket by a pressure so light that the ball can be moved in any direction in the operation of leveling the compass.

The ball is placed either in the brass head of the Jacob Staff already shown with the previous instrument, or still better, in the compass tripod seen in the engraving of the Vernier Transit beyond.

The superiority of the vernier over the plain compass consists in its adaptation to the retracing the lines of an old survey, and to the surveys of the U. S public lands, where the lines are based on a true meridian.

Variation of the Needle.

It is well known that the magnetic needle, in almost all parts of the United States, points more or less to the east or west of a true meridian, or north and south line.

This deviation, which is called the VARIATION OR DECLINATION of the needle, is not constant, but increases or decreases to a very sensible amount in a series of years.

Thus at Troy. N. Y., a line bearing in 1838, N. 31° E., would in 1874, with the same needle, have a bearing of about N. 32° E., the needle having thus in that interval travelled a full degree to the west.

For this reason, therefore, in running over the lines of a farm from field notes of some years standing, the surveyor would be obliged to make an allowance, both perplexing and uncertain, in the bearing of every line.

To avoid this difficulty the *vernier* was devised, the arrangement of which we shall now describe.

THE VERNIER is divided on its edge to thirty equal parts, and figured in two series on each side of the centre line.

In the same plane with the vernier is an arc or limb, fixed to the main plate of the compass, and graduated to half degrees.

The surfaces of both vernier and limb are silvered.

On the vernier are thirty equal divisions, which exactly correspond in length with thirty-one of the half degrees of the limb.

Each division of the vernier is, therefore, one-thirtieth or, in other words, one minute longer than a single division of the limb.

To READ THE VERNIER.—In "reading" the vernier, if it is

moved to the right, count the minutes from its zero point to the left, and vice versa. Proceed thus until a division on the vernier is found exactly in line with another on the limb, and the lower row of figures on the vernier will give the number of minutes passed over. When the vernier is moved more than fifteen minutes to either side the number of the additional minutes up to thirty or one-half degree of the limb is given by the upper row of figures on the opposite side of the vernier.

To read beyond thirty, add the minutes given by the vernier to that number, and the sum will be the correct reading.

In all cases when the zero point of the vernier passes a whole degree of the limb, this must be added to the minutes, in order to define the distance over which the vernier has been moved.

To TURN OFF THE VARIATION.—It will now be seen that the surveyor having the vernier compass, can by moving the vernier to either side, and with it of course the compass circle attached, set the compass to any variation.

He therefore places his instrument on some well defined line of the old survey, and turns the tangent screw until the needle of his compass indicates the same bearing as that given in the old field notes of the original survey.

Then screwing up the clamping nut underneath the vernier, he can run all the other lines from the old field notes without further alteration.

The reading of the vernier on the limb in such a case would give the change of variation at the two different periods.

The variation of the needle at any place being known, a true meridian, or north and south line, may be run by moving the vernier to either side, as the variation is east or west, until the arc passed over on the limb is equal to the angle of variation; and then turning the compass until the needle is made to cut the zeros on the divided circle, when the line of the sights would give the direction of the true meridian of the place.

Such a change in the position of the vernier is necessary in surveying the U. S. public lands, which are always run from the true meridian.

"THE LINE OF NO VARIATION, as it is called, or that upon which the needle will indicate a true north and south direction, is situated in the United States, nearly in an imaginary line drawn from the middle of lake Erie to Cape Hatteras, on the coast of North Carolina.

A compass needle, therefore, placed east of this line would have a variation to the west, and when placed west of the line, the variation would be to the east, and in both cases the variation would increase as the needle was carried farther from the line of no variation.

Thus in Minnesota the variation is from 15° to 16° to the east, while in Maine it is from 17° to 18° to the west.

At Troy, in the present year, 1874, the variation is about 9° to the west, and is increasing in the same direction from two to three minutes annually.

TO READ TO MINUTES.—A less important use of the vernier is to give a reading of the needle to single minutes, which is obtained as follows:

First be sure, as in all observations, that the zero of the vernier exactly corresponds with that of the limb ; then noting the number of whole degrees given by the needle, move back the compass circle with the tangent screw until the nearest whole degree mark is made to coincide with the point of the needle, read the vernier as before described, and this reading added to the whole degrees will give the bearing to minutes.

To use the Vernier Compass.

Proceed in the same manner as directed in regard to the Plain Compass, when making new surveys, always taking care that the vernier is set at zero and securely clamped by screwing up the nut beneath the plate.

In surveying old farms, allowance and correction must be made for the variation, as just described.

Sizes of the Vernier Compass.

We make three sizes of this compass, having needles of four, five and six inches long respectively, the main plates of the two largest being over fifteen inches long; and of the smallest size, thirteen inches, the sights of the last are also about an inch shorter.

In the four and five inch Vernier Compasses, the variation arc is within the compass circle like that of the railroad compass hereafter described, and the variation is set off to minutes by a pinion head underneath the plate; the circle is also clamped at any variation by a screw placed opposite the pinion.

Weight of the Vernier Compasses.

The average weights of the different sizes, including the brass head of the Jacob Staff, beginning with the smallest, are respectively $5\frac{1}{2}$, $7\frac{1}{2}$ and $9\frac{1}{2}$ pounds.

Surveying Instruments.

THE VERNIER TRANSIT.

This instrument, shown in the engraving opposite, resembles the Vernier Compass in its construction and general principles, differing from it mainly in the use of a telescope in place of the ordinary sights. The variation of the needle is set off by a pinion, as shown, and the circle clamped by a nut underneath the plate as usual.

The instrument is clamped to the spindle and secured from falling from it by the clamp screw and spring catch, seen on opposite sides of the socket in the engraving.

The Vernier Transit should always be used with the compass tripod at least, as shown in the engraving, and often, especially when furnished with the extra attachments to telescope, is placed upon the light leveling tripod, shown with the Surveyor's Transit beyond, and described in our account of that instrument.

The needle of this instrument is either four, five or six inches long, as the surveyor may prefer, the one with the six inch needle being shown in the engraving, and generally selected by our customers.

The advantages of the Vernier Transit over the ordinary compass, are due mainly to the telescope and its attachments which we shall proceed to describe in detail.

VERNIER TRANSIT.

Price, as shown above $ 108,00.

Made by

W. & L. E. GURLEY,

TROY, N.Y.

BENJ. D. BENSON, N.Y.

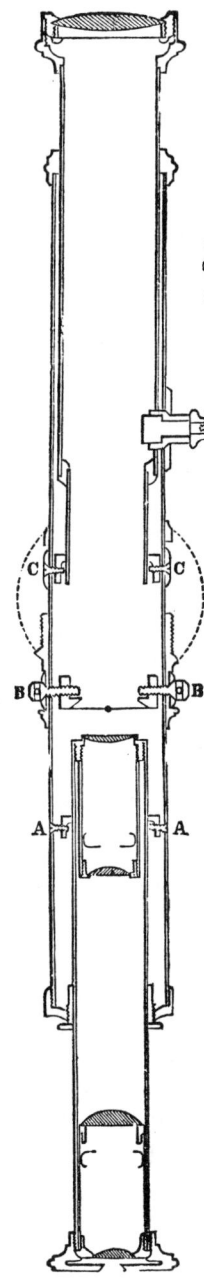

Fig. 4.

The telescope is from ten to twelve inch‑ es long, and sufficiently powerful to see and set a flag at a distance of two miles in a clear day.

The cross-bar in which it is fixed, turns readily in the standards, so that the tele‑ scope can be turned in either direction, and back and fore sights be taken without moving the instrument.

Like all telescopes used by us in our in‑ struments, it shows objects in an erect position.

THE TELESCOPE.—The interior construc‑ tion of the telescope of the Vernier Transit, which is very similar to those of the other instruments we shall describe, is well shown in the longitudinal section represen‑ ted in fig. 4.

As here seen, the telescope consists es‑ sentially of an object-glass, an eye-piece tube, and a cross-wire ring or diaphragm.

The object-glass is composed of two lenses, one of flint, the other of crown glass, which are so made and disposed as to show the object seen through it without color or distortion.

The object glass and the whole telescope is therefore said to be " achromatic."

The eye-piece is made up of four plano‑ convex lenses, which, beginning at the eye end, and proceeding on, are called respec‑ tively, the eye, the field, the amplifying, and the object lenses.

Together, they form a compound microscope, magnifying the minute image of any object formed at the cross-wires by the interposition of the object glass.

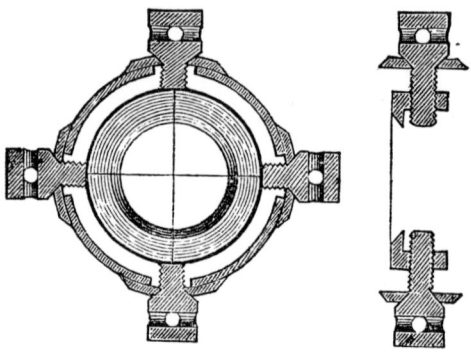

Fig. 5.

THE CROSS WIRES.—The cross-wire diaphragm, two views of which are here exhibited, is a small ring of brass, sus. ended in the tube of the telescope by four capstan head screws, which press upon the washers shown on the outside of the tube.

The ring can thus be moved in either direction by working the screws with an ordinary adjusting pin.

Across the flat surface of the ring two fine fibres of spider's web are extended at right angles to each other, their ends being cemented with beeswax or varnish, into fine lines cut in the metal of the ring.

The intersection of the wires forms a very minute point, which, when they are adjusted, determines the optical axis of the telescope, and enables the surveyor to fix it upon an object with the greatest precision.

The imaginary line passing through the optical axis of the telescope, is termed the "line of collimation," and the operation of bringing the intersection of the wires into the optical axis, is called the " adjustment of the line of collimation." This will be hereafter described.

The openings in the telescope tube are made considerably larger than the screws, so that when these are loosened, the whole ring can be turned around for a short distance in either direction.

The object of this will be seen more plainly, when we describe the means by whicn the wire is made truly vertical.

The sectional view of the telescope (fig. 4) also shows two moveable rings, one placed at A A, the other at C C, which are respectively used, to effect the centering of the eye-piece, and the adjustment of the object-glass slide.

The centering of the eye-tube is performed after the wires have been adjusted, and is effected by moving the ring, by means of the screws, shown on the outside of the tube, until the intersection of the wires is brought into the centre of the field of view.

The adjustment of the object slide, which will be fully described in our account of the Leveling Instrument, secures the movement of the object-glass in a straight line, and thus κeeps the line of collimation in adjustment through the whole range of the slide, preventing at the same time what is termed the " travelling " of the wires.

This adjustment, which is peculiar to our telescopes, is always made in the process of construction, and needing no further attention at the hands of the engineer, is concealed within the hollow ball of the telescope axis.

Optical Principles of the Telescope.

In order that the advantages gained by the use of the tele scope may be more fully understood, we shall here venture briefly to ccnsider the optical principles involved in its con struction.

We are said to " see " objects because the rays of light which proceed from all their parts, after passing through the pupil of the eye, are by the crystalline lens and vitreous humor, converged to a focus on the retina, where they form a very minute inverted image; an impression of which is conveyed to the brain by the optic nerve

The rays proceeding from the extremities of an object, and crossing at the optic center of the eye, form the " visual angle," or that under which the object is seen.

The apparent magnitude of objects depends on the size of the visual angle which they subtend, and this being great or small, as the object is near or distant—the objects will appear large or small, in an inverse proportion to the distances which separate them from the observer.

Fig. 6.

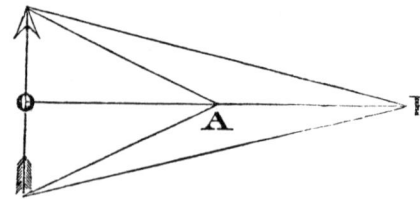

Thus, (in fig. 6,) if the distance O A is one-half of O B, the visual angle, subtended by the object at the point A, and therefore the apparent magnitude of the object will be twice that observed at B.   If, therefore, the visual angle subtended by any object, can be made by any means twice as large, the same effect will be produced as if the observer were moved up over one half the intervening distance.

Now this is the principal advantage gained in the use of a telescope.

The object-glass receiving the rays of light which proceed from all the points of a visible object, converges them to a focus at the cross-wires, and there forms a minute, inverted, and very bright image, which may be seen by placing a piece of ground glass to receive it at that point.

The eye-piece acting as a compound microscope, magnifies

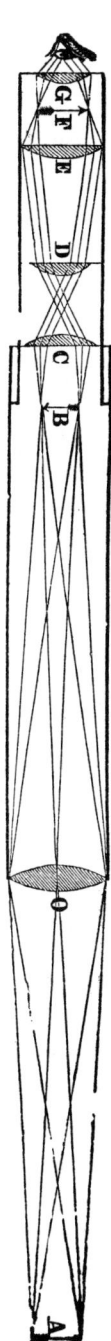

this image, restores it to its natural position, and conveys it to the eye.

The visual angle which the image there subtends, is as many times greater than that which would be formed without the use of the telescope, as the number which expresses its magnifying power.

Thus, a telescope which magnifies twenty times, increases the visual angle just as much, and therefore diminishes the apparent distance of the object twenty times—or in other words, it will show an object two hundred feet distant, with the same distinctness as if it was distant only ten feet from the naked eye.

The accompanying cut, (fig. 7) which we are kindly permitted to copy from an excellent treatise on surveying, by Prof. Gillespie of Union College, will give a correct idea of the manner in which the rays of light coming from an object are affected, by passing through the several glasses of a telescope.

We shall only consider the rays which proceed from the extremities; these, after passing through the object-glass, here shown as a single lens, are conveyed to the point B, the centre of the cross-wires and the common focus of the object and eye-glasses. At this place the rays cross each other and the image is inverted.

The rays next come to the object lens C, and passing through it are refracted so as again to cross each other, and come thus to the amplifying lens D. By this they are again refracted, made more nearly parallel, and thus reach the large field lens E. After passing through this, the form a magnified and erect image in the focus of the eye lens G. By the eye lens the image is still further magnified, and at last enters the eye of the observer, subtending an angle as much greater than that at the point O, as is the magnifying power of the telescope.

In place of the eye-piece of four lenses, which we have just been considering, and which is exclusively used in all American instruments made at the present day; another, which has but three lenses, is often seen in the telescopes of imported instruments.

This latter, which inverts the object, though saving a little more light than the former, is exceedingly troublesome to the inexperienced observer, and has never been popular in American engineering.

To ascertain the Magnifying Power of a Telescope.

Set up the instrument about twenty or thirty feet from the side of a white wooden house, and observe through the telescope the space covered by one of the boards in the field of the glass; then still keeping that eye on the telescope, hold open the other with the finger, if necessary, and look with it at the same object. By steady and careful observation there will appear on the surface of the magnified board, a number of smaller ones seen by the naked eye, count these, and we shall obtain the magnifying power.

If the limits of the magnified board, as seen through the telescope, can be noted so as to be remembered after the eye is removed, the number of boards contained in this space may then be easily counted.

The side of an unpainted brick wall, or any other surface containing a number of small, well marked and equal objects, may be observed, in place of the surface we have described

The operation described requires great care and close

observation, but may be performed with facility after a little practice.

We have spoken of the effect of the telescope in magnifying objects, but have not mentioned what is termed its " illuminating power."

This arises from the great diameter or aperture of the object-glass compared with that of the pupil of the eye, which enables the observer to intercept many more rays of light, and bring the object to the eye highly illuminated.

The advantage gained in this increase of light,depends, as is evident, on the size of the object glass, and the perfection with which the lenses transmit the light without absorbing or reflecting it.

The superficial magnifying power of a telescope, is found oy squaring the number which expresses its linear magnifying power; thus a telescope which magnifies twenty times, increases the surface of an object four hundred times.

Before an observation is made with the telescope, the eyepiece should be moved in or out, until the wires appear distinct to the eye of the operator; the object-glass is then adjusted by turning the pinion head until the object is seen clear and well defined, and the wires appear as if fastened to its surface.

The intersection of the wires, being the means by which the optical axis of the telescope is defined, should be brought precisely upon the centre of the object to which the instrument is directed.

Having thus briefly considered the principles, we shall now proceed to describe the

Attachments of the Telescope.

A telescope is said to be " plain " when it is without any appendages to its tube or axis, as that of the Engineer's

Transit shown in the engraving, and most instruments are made in that manner.

Many surveyors, however, prefer to add these conveniences, and we shall now consider them in detail.

The Clamp and Tangent—Consists of a ring which encircles one end of the telescope axis, and has also an arm projecting below, for the attachment of a tangent screw. The ring is securely fastened at will to the axis by a clamp screw inserted on one side, and the telescope then moved slowly up or down by turning the tangent screw.

The clamp and tangent ought always to accompany the vertical circle, and the level on the telescope.

Vertical Circle.—A divided circle as seen in the cut of the Vernier Transit, is often attached to the axis of the telescope, giving, with a vernier, the means of measuring vertical angles with great facility.

We make two sizes of these circles, one of 3½ inches diameter, seen with this instrument, the other an inch larger, and shown in the cut of the Surveyor's Transit. The former is graduated to single degrees, and reads by the vernier, to five minutes of a degree. The latter, divided to half degrees, gives a reading, with the vernier, to single minutes.

The vertical circle is fitted firmly to the telescope axis, and fastened with a screw, so that it remains permanent.

The vernier, however, may be shifted in either direction, by loosening the screws which confine it to the standards.

The vernier of the small circle is divided into twelve equal parts, which correspond with thirteen degrees on the circle.

Each division of the vernier is, therefore, one-twelfth of one degree, or five minutes longer than a single division of the circle, so that the angles are read to five minutes of a degree

The vernier is double, having its zero poi1t in the middle, and the reading up to thirty minutes, is said to be direct; that is, if the circle is moved to the right, the minutes are read off on the right side of the vernier, and vice versa.

The minutes beyond thirty are obtained on the opposite side, and in the lower row of figures.

By following these directions, and noticing the first divisions on the circle and vernier, which exactly correspond, the surveyor can obtain a reading to five minutes with great facility.

LEVEL ON TELESCOPE.—Besides the vertical circle, there is sometimes a small level attached to the telescope of this and other instruments, which we shall hereafter describe.

Such an attachment is shown in the cut of the Surveyor's Transit, and its adjustment and advantages will be explained in our account of that instrument.

SIGHTS ON TELESCOPE.—We are sometimes desired by surveyors to place a pair of short sights on the upper side of the telescope tube.

They are best made to fold close to the tube when not in use, like those of the pocket compass, described hereafter.

These sights are useful in taking back sights without turning the telescope, and in sighting through bushes or in the forest, and as the telescope can be turned up or down, answer all the purposes of the longer sights of the ordinary compass.

SIGHTS FOR RIGHT ANGLES.—Besides the sights just mentioned, we have often attached others to the plate of the instrument, on either side of the compass circle or on the standards.

These being adjusted to the telescope give a very ready means of laying off right angles, or running out offsets, without changing the position of the instrument

To adjust the Vernier Transit.

THE LEVELS of this instrument have a capstan head screw at each end, and are adjusted with a steel pin in the same manner as those of the Plain compass.

THE NEEDLE is also adjusted as described in our account of that instrument.

LINE OF COLLIMATION.—To make this adjustment, which is, in other words, to bring the intersection of the wires into the optical axis of the telescope, so that the instrument, when placed in the middle of a straight line will, by the revolution of the telescope, cut its extremities—proceed as follows:

Set the instrument firmly on the ground and level it carefully; and then having brought the wires into the focus of the eye-piece, adjust the object-glass on some well defined point, as the edge of a chimney or other object, at a distance of from two to five hundred feet; determine if the vertical wire is plumb, by clamping the instrument firmly to the spindle and applying the wire to the vertical edge of a building, or observing if it will move parallel to a point taken a little to one side; should any deviation be manifested, loosen the cross-wire screws, and by the pressure of the hand on the head outside the tube, move the ring around until the error is corrected.

The wires being thus made respectively horizontal and vertical, fix their point of intersection on the object selected ; clamp the instrument to the spindle, and having revolved the telescope, find or place some good object in the opposite direction, and at about the same distance from the instrument as the first object assumed.

Great care should always be taken in turning the telescope, that the position of the instrument upon the spindle is not in the slightest degree disturbed.

Now, having found or placed an object which the vertical wire bisects, unclamp the instrument, turn it half way around, and direct the telescope to the first object selected; having bisected this with the wires, again clamp the instrument, revolve the telescope, and note if the vertical wire bisects the second object observed.

Should this happen, it will indicate that the wires are in adjustment, and the points bisected are with the centre of the instrument, in the same straight line.

If not, however, the space which separates the wires from the second point observed, will be double the deviation of that point from a true straight line, which may be conceived as drawn through the first point and the centre of the instrument, since the error is the result of two observations, made with the wires when they are out of the optical axis of the telescope.

Fig. 8.

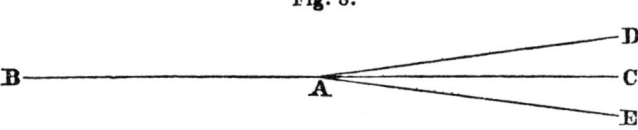

For as in the diagram, let A represent the centre of the instrument, and B C the imaginary straight line, upon the extremities of which the line of collimation is to be adjusted.

B represents the object first selected, and D the point which the wires bisected, when the telescope was made to revolve.

When the instrument is turned half around, and the telescope again directed to B, and once more revolved, the wires will bisect an object, E, situated as far to one side of the true line as the point D is on the other side.

The space, D E, is therefore the sum of two deviations of the wires from a true straight line, and the error is made very apparent

In order to correct it, use the two capstan head screws on the sides of the telescope, these being the ones which affect the position of the vertical wire.

Remember that the eye-piece inverts the position of the wires, and therefore that in loosening one of the screws and tightening the other on the opposite side, the operator must proceed as if to increase the error observed. Having in this manner moved back the vertical wire until, by estimation, one-quarter of the space, D E, has been passed over, return the instrument to the point B, revolve the telescope, and if the correction has been carefully made, the wires will now bisect a point, C, situated midway between D and E, and in the prolongation of the imaginary line, passing through the point B and the centre of the instrument.

To ascertain if such is the case, turn the instrument half around, fix the telescope upon B, clamp to the spindle, and again revolve the telescope towards C. If the wires again bisect it, it will prove that they are in adjustment, and that the points, B, A, C, all lie in the same straight line.

Should the vertical wire strike to one side of C, the error must be corrected precisely as above described, until it is entirely removed.

Another method of adjusting the line of collimation often employed in situations where no good points in opposite directions can be selected upon which to reverse the wires, may here be described.

The operator sets up the instrument in some position which commands a long sight in the same direction, and having leveled his instrument, clamps to the spindle, and with the telescope locates three points which we will term A, B and C, which are distant from the instrument about one, two and three hundred feet respectively.

These points, which are usually determined by driving a nail into a wooden stake set firmly into the ground, will all be in the same straight line, however much the wires are out of adjustment, since the position of the instrument remains unchanged during the whole operation.

Having fixed these points he now moves the instrument to B, and sets its centre directly over the nail head, by letting lown upon it the point of a plumb-bob suspended from the tripod.

Then having leveled the instrument, he directs the wires to A, clamps to the spindle and revolves the telescope towards C. Should the wires strike the nail at that point, it would show that they were in adjustment.

Should any deviation be observed, the operator must correct it by moving the wire with the screws until, by estimation, half the error is removed.

Then bringing the telescope again upon either A or C, and revolving it, he will find that the wires will strike the point in the opposite direction if the proper correction has been applied.

If not, repeat the operation until the telescope will exactly cut the two opposite points, when the intersection of the wires will be in the optical axis, and the line of collimation in adjustment.

In our description of the previous operation, we have spoken more particularly of the vertical wire, because in a revolving telescope this occupies the most important place, the horizontal one being employed mainly to define the centre of the vertical wire, so that it may be moved either up or down without materially disturbing the line of collimation.

The wires being adjusted, their intersection may now be brought into the centre of the field of view by moving the screws A A, shown in the sectional view of the telescope, Fig. 4, which are slackened and tightened in pairs, the movement being now direct, until the wires are seen in their proper position.

It is here proper to observe, that the position of the line of collimation depends upon that of the object-glass, solely, so that the eye-piece may, as in the case just described, be moved in any direction, or even entirely removed and a new one substituted, without at all deranging the adjustment of the wires.

THE STANDARDS.—In order that the wires may trace a vertical line as the telescope is moved up or down, it is necessary that both the standards of the telescope should be of precisely the same height.

To ascertain this and make the correction if needed, proceed as follows:

Having the line of collimation previously adjusted, set the instrument in a position where points of observation, such as the point and base of a lofty spire, can be selected, giving a long range in a vertical direction.

Level the instrument, fix the wires on the top of the object and clamp to the spindle; then bring the telescope down, until the wires bisect some good point, either found or marked at the base; turn the instrument half around, fix the wires on the lower point, clamp to the spindle, and raise the telescope to the highest object.

If the wires bisect it, the vertical adjustment is effected; if they are thrown to either side this would prove that the standard opposite that side was the highest, the apparent error being double that actually due to this cause.

To correct it, we now make one of the bearings of the axis movable, so that by turning a screw underneath this sliding piece, as well as the screws which hold on the cap of the standard, the adjustment is made with the utmost precision.

This arrangement, which is common to all our telescope instruments, is very substantial and easily managed.

THE VERTICAL CIRCLE.—When this attachment requires adjustment, proceed by leveling the instrument carefully, and having brought into line the zeros of the wheel and vernier, find or place some well defined point or line which is cut by the horizontal wire.

Turn the instrument half around, revolve the telescope, and fixing the wire upon the same point as before, note if the zeros are again in line.

If not, loosen the screws and move the zero of the vernier over half the error; bring the zeros again into coincidence, and proceed precisely as at first described until the error is entirely corrected, when the adjustment will be completed.

Should it be desired, at any time, the circle can be removed by the surveyor and replaced at pleasure.

THE LEVEL ON TELESCOPE.—The adjustment of this will be best considered when we come to speak of the Surveyors' Transit.

ADJUSTMENTS IN GENERAL.—We ought here to say that the above adjustments, as well as all the others which we have previously explained or may hereafter describe, are always made by us in person, but are given in this work in order that the surveyor and engineer may fully understand their instruments, and be enabled to detect and remedy errors and accidents, which in practice will often occur.

To use the Vernier Transit.

This instrument is used on the ordinary ball and spindle, placed most commonly in the compass tripod, as shown in the engraving.

LEVELING HEAD.—Sometimes leveling screws with the parallel plates, and which together we shall designate the "leveling head," with a clamp and tangent movement, are used with this instrument as well as with the Surveyor's Transit.

This leveling head can be unscrewed from the legs, and is packed in the instrument box; it is of very moderate cost, and in almost every situation is infinitely superior to a ball and socket or any other support.

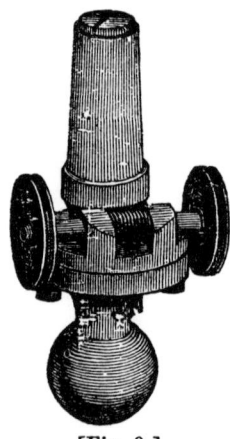

COMPOUND BALL.—We also manufacture what may be termed a "compound ball spindle," which has a tangent movement, and gives all the perfection of more costly arrangements, with a very moderate expense.

As represented in the cut, it has an interior spindle, around which an outside hollow cylinder is moved by turning the double-headed tangent screw, which has in the middle an endless screw, working into teeth cut spirally around in a groove

[Fig. 9.]   of the cylinder. The compass, or other instrument, revolves on the outside socket, precisely as if placed on a common ball spindle; but when a slower movement is required, can be made fast by the clamp screw, and then turned gradually around the interior spindle by the tangent screw, until the slote of the sight or the intersection of the wires, is made to bisect the object with the utmost certainty.

The compound ball may be placed either in a jacob-staff socket or compass tripod.

LEVELING SOCKET. — A convenient arrangement for use, either with this instrument or with a sight compass, is shown in the leveling socket described in our account of the solar compass beyond.

The socket may be used either with the ordinary compass ball or the compound ball, as there represented, and gives a very rapid and accurate means of leveling the instrument.

THE SPRING CATCH, described in our account of the Plain Compass, is always attached to the socket of this instrument, whether placed upon a ball or tripod, so that it cannot slip off from the spindle in carrying.

THE CLAMP SCREW, in the side of the socket of this instru

ment, is shown in Fig 3, and by pressing a brass spring in the interior against the spindle, serves to fix the instrument in any position.

THE VERNIER is moved by the pinion head, now always placed beneath the plate, precisely as described in our account of the Vernier Compass, and is read to minutes in the same manner.

There is also a clamp nut underneath the circle, by which it is securely fixed in any position, which must be loosened whenever the vernier is moved by the pinion.

THE NEEDLE LIFTING SCREW is the same as those of the compasses previously described.

IN SURVEYING with this instrument the operator proceeds precisely as with the Vernier Compass, keeping the south end towards his person, reading the bearings of lines from the north end of the needle, and using the telescope in place of sights, revolving it as objects are selected in opposite directions.

PARALLAX.—Before an observation is made with the telescope, the eye-piece should be moved in or out until the wires appear distinct to the eye of the operator, the object-glass may then be placed in position by turning the pinion head on the top of the telescope, until the object is seen clear and well defined, and the wires appear as if fastened to its surface.

When, on the contrary, the wires are not perfectly distinct, the observer, by moving his eye to either side of the small aperture of the eye piece, will cause the wires to "travel" on the object, and thus occasion what is termed the "error of parallax."

The intersection of the wires being the means by which the optical axis of the telescope is defined, should be brought precisely upon the centre of the object to which the instrument is directed.

To TAKE ANGLES OF ELEVATION.—Level the instrument carefully, fix the zeros of the circle and vernier in line, and note
the height cut upon the staff or other object, by the horizontal wire; then carry the staff up the elevation, fix the wire
again upon the same point, and the angle of elevation will
be read off by the vernier.

By careful usage, the adjustments of the Vernier Transit
will remain as permanent as those of the ordinary compass
the only one liable to derangement being that of the line of
collimation.

This should be examined occasionally, and corrected in the
manner previously described.

Repairs of the Vernier Transit.

These being in great part already spoken of, it will be
necessary to consider only such as belong to the telescope.

To REPLACE THE CROSS WIRES.—Take out the eye-piece
tube, together with the little ring by which it is centered,
and having removed two opposite cross-wire screws, with
the others turn the ring until one of the screw holes is
brought into view from the open end of the telescope tube,
in this thrust a stout splinter of wood or a small wire so as
to hold the ring when the remaining screws are withdrawn ;
the ring is then taken out and is ready for the wires.

For these the web of the spider is to be preferred above
any thing else, but when this is not obtainable, a fine silk
fibre may be substituted.

We usually procure our webs from the living manufacturer
directly, selecting those of a yellowish-brown color, as furnishing the most perfect product.

The spider being held between the thumb and finger of an
assistant, in such position as to suffer no serious injury, and
at the same time be unable to make any effectual resistance
with his extremities, the little fibre may be drawn out at
pleasure, and being placed in the fine lines cut on the surface

of the diaphragm, is then firmly cemented to its place by applying softened beeswax with the point of a knife blade

In case the spider is not procurable, a fine strand of a web which is free from dust, and long enough to serve for both wires, may be selected.

In such times as the spiders remain in their winter quarters, we have been able to procure very good fibres from a box in which a number had been confined.

When the wires are cemented, the ring is returned to its position in the tube, and either pair of screws being inserted, the splinter or wire is removed, and the ring turned until the other screws can be replaced.

Care must also be taken that the same side of the ring is turned to the eye-piece as before it was removed.

When this has been done, the eye-tube is inserted, and its centering ring brought into such a position that the screws in it can be replaced, and then by screwing on the end of the telescope, the little cover into which the eye-tube is fixed, the operation will be completed.

To CLEAN THE TELESCOPE.—The only glasses that will ordinarily require cleaning, are the object-glass on its outside surface, and the little eye-lens, which is exposed when the cap of the eye-tube is removed.

To remove the dust from these use a very soft and clean silk or cotton cloth, and be careful not to rub the same part of the cloth a second time on the surface of the glass.

No one should ever be allowed to touch the glasses with the fingers or with a dusty cloth.

Excellencies of the Vernier Transit.

These are due chiefly to the telescope and its attachments, and from what has already been said, it will appear are such as to render this instrument greatly superior to one provided with the ordinary sights.

1. The magnifying power of the telescope enables the sur-

veyor to take accurate observations at distances entirely beyond the reach of the naked eye.

2. The fine intersection of the cross-wires can be set precisely upon the centre of the object.

3 The revolving property of the telescope gives the means of running long lines up or down steep ascents or descents with perfect ease, where, with the short sights of the ordinary compass, two or three observations would have to be taken.

4. The use of a telescope entirely avoids the incessant trying of the eyes, experienced in surveys with the ordinary sights.

5. With the telescope, lines can be run through the forest or brushwood, and the flagstaff distinguished with much greater certainty than through the sights of a compass.

This statement may appear very unreasonable to those not familiar with the instrument, and these, in fact, raise the greatest objection to a telescope, from its supposed unfitness for surveys in such locations.

They have only to use it a few times in this kind of work, in connection with a flagstaff, painted white or covered with paper, to distinguish it from the surrounding objects, to be convinced of its great superiority.

In the Vernier Transit, as furnished by us, is supplied, as we believe, to the surveyor the most perfect of all needle instruments, and this at a cost but little above that charged by other makers for a sight compass.

The advantages of the telescope and its attachments are so great that a surveyor, accustomed to them, would find it difficult to content himself with the ordinary compass, and such in fact is the universal testimony of those familiar with the Vernier Transit.

Sizes of the Vernier Transit

We make three sizes, having respectively needles of four five and six inches long.

The telescopes of our five and six inch transits, are both eleven inches long, and reverse at either end ; the telescope of the four inch size is about seven inches, and the whole instru· ment very light and portable.

Weights of the Vernier Transits.

The average weights of the different sizes, not including the tripods are, for the four inch instrument. five pounds ; the five inch, eight and a half, and the six inch, eleven pounds.

Needle Instruments.

We have now described the instruments included under the division termed Needle Instruments, in the beginning of this work.

As there stated, the Plain and Vernier Compasses and the Vernier Transit depend for their accuracy and value, mainly upon the perfection of movement of the magnetic needle.

With such instruments, the greater part of the surveying in our country has been, and will for a long time in the future, continue to be done.

And though with the improvements made in these instruments, a good surveyor may, with great care and skill, do work with a surprising degree of accuracy and perfection, yet all needles are liable to many irregularities.

Imperfections of the Needle.

These may arise either from the loss of magnetic virtue in the poles, the blunting of the centre-pin, or the attraction exerted upon it by bodies of iron, whose presence may be entirely unsuspected.

The two first of these errors may be easily remedied in the manner we have described.

LOCAL ATTRACTION.—The third and most frequent source of inaccuracy, may be detected by taking back sights, as well as fore sights, upon every line run with the needle, and by the agreement of the bearings, determining the true direction of the line.

Sometimes a compass may have little particles of iron concealed within the surface of the metal circle or plates

It is the business of the maker to examine every instrument, in search of this defect, by trying the reversion of the needle upon all points of the divided circle.

If the needle should fail to reverse, when the compass is turned half around, and the sights directed a second time upon any object, the instrument should be thrown aside and never sold.

Besides the dificulties caused by the above imperfections, the variation of the needle is a frequent source of annoyance.

What is termed the secular variation, we have already mentioned in our acccount of the Vernier Compass, we will now speak of the

DIURNAL VARIATION.—This is owing to the influence of the sun, which, in summer, will cause the needle to vary from ten to fifteen minutes in a few hours, when .exposed to its fullest influence.

To guard against these causes of inaccuracy in the use of needle instruments, the surveyor will need the greatest care and attention ; and yet, with all the precautions than can be suggested, the difficulty of measuring horizontal angles with certainty, and to a sufficient degree of minuteness by the needle alone, has caused a demand to be felt more and more sensibly in all parts of the country for instruments, in the use of which the surveyor may proceed with assured accuracy and precision.

Indeed, in Canada, so great is the distrust of needle instruments, that the Provincial Land Surveyors are forbidden to use an instrument in their land surveys, unless it is capable of taking angles independently of the needle.

To supply the demand thus created for increased perfection in the implements of the surveyor, we manufacture a variety of instruments ; three of which we shall now describe under the names of *The Railroad Compass, The Surveyor's Transit* and the *Solar Compass.*

Surveying Instruments.

THE RAILROAD COMPASS.
Fig. 10.

As shown in Fig. 10, this instrument has the main plate, levels, sights, and needle of the ordinary instrument, and has also underneath the main plate a divided circle or limb by which horizontal angles to single minutes can be taken independently of the needle.

The verniers are attached to the under surface of the main plate the openings through which they are seen being covered with slips of glass to protect the divisions from dust and moisture; only one of the verniers is shown in the cut.

The connection between the two plates is made by a clamp and tangent movement shown at e, by which they can be fas·

tened together or released at will, or moved slowly around each other as may be desired in the use of the compass.

The needle lifting screw is shown near the clamp screw, on the same end of the plate.

On the opposite side of the compass circle is seen the head *a* of a pinion working into a circular rack fixed to the edge of the compass circle, and thus enabling the surveyor to move the compass circle about its centre in setting off the variation of the needle, precisely as in the case of the vernier compass.

The variation is read to single minutes by a vernier and divided arc, partially shown near the letter S in the cut.

Near the pinion head is also shown a clamp screw, by which the circle is securely fixed when moved to the proper position.

The sockets upon which the plates of this instrument turn are long and well fitted, and the movement of the vernier plate around the limb is almost perfectly free from friction.

THE GRADUATED CIRCLE or limb is divided to half degrees, and figured in two rows, viz : from 0° to 90°, and from 0° to 360°; sometimes but a single series is used, and then the figures run from 0° to 360°, or from 0° to 180° on each side.

The figuring, which is the same upon this as in the other angular instruments we shall hereafter describe, is varied when desired by the surveyor. The first method is our usual practice.

THE VERNIERS are double, having on each side of the zero mark thirty equal divisions corresponding precisely with twenty-nine half degrees of the limb; they thus read to single minutes, and the number passed over is counted in the same direction in which the vernier is moved.

The use of two opposite verniers in this and other instru-ments gives the means of "cross questioning" the gradua-tions, the perfection with which they are centered and the dependence which can be placed upon the accuracy of the angles indicated.

THE NEEDLE of this instrument is five or five and a half inches long, and made precisely like those previously described.

THE ADJUSTMENTS of this instrument, with which the surveyor will have to do, have been already described.

To use the Railroad Compass.

It can be set upon the common compass ball, or still better, the tangent ball already described, placed either in a jacob-staff socket, a compass tripod, or the leveling socket and tripod as shown with the solar compass.

We have also adapted to many of these instruments, the leveling tripod head, with clamp and tangent movement, and this is preferable to any other support.

To TAKE HORIZONTAL ANGLES.—First level the plate and set the limb at zero, fix the sights upon one of the objects selected, and clamping the whole instrument firmly to the spindle, unclamp the vernier plate and turn it with the hand, until the sights are brought nearly upon the second object; then clamp to the limb, and with the tangent screw fix them precisely upon it.

The number of degrees and minutes read off by the vernier, will give the angle between the two objects, taken from the centre of the instrument.

It will be understood that the horizontal angles can be taken in any position of the verniers, with reference to the zero point of the limb; we have given that above as being the usual method and liable to the fewest errors.

It is advisable where great accuracy is required, in this and other instruments furnished with two verniers, to obtain the readings of the limb from both, add the two together and halve their sum; the result will be the mean of the two readings, and the true angle between the points observed.

Such a course is especially necessary when the readings of the verniers essentially disagree, as may sometimes happen when the instrument has been injured by an accident.

USE OF THE NEEDLE.—In taking horizontal angles as just described, the magnetic bearings of the two objects are often noted, and thus two separate readings of the same angle, one by the limb, the other by the needle, are obtained, to be used as checks upon each other to prevent mistakes.

TO TURN OFF THE VARIATION OF THE NEEDLE.—Having leveled the instrument, set the limb at zero, and place the sights upon the old line, note the reading of the needle, and make it agree with that given in the field notes of the former survey, by turning the compass circle about its centre by the pinion a.

Now, clamp the compass circle firmly by the clamp screw, and the number of degrees or minutes passed over by the vernier of the compass circle will be the change of variation in the interval between the two surveys.

TO SURVEY with this instrument, the operator should turn the south side of the compass face towards his person, and having brought the zeros of the limb and vernier plate in contact, clamp them, and proceed as directed in our account of the Plain Compass.

Of course it will be understood that lines can be run and angles measured by the divided limb and verniers, entirely independent of the needle, which, in localities where local attraction is manifested, is very serviceable.

The accuracy and minuteness of horizontal angles indicated by this instrument, together with its perfect adaptation to all the purposes to which the Vernier Compass can be applied, have brought it into use in many localities, where the land is so valuable as to require more careful surveys than are practicable with a needle instrument.

Single Vernier Railroad Compass.

We have just introduced a new style of this instrument, essentially alike that already described, but of somewhat simpler construction in its sockets, and having but a single vernier to the limb.

This new instrument, though afforded at a price materially lower than the other, is still in every way accurate and reliable.

Size and Weight of the Railroad Compass, Single Vernier.

We make two sizes of this instrument, viz.: five, and five and a half inch needle; the largest size, including the brass head of the jacob staff, weighing ten and a half pounds, and the five inch, ten pounds.

Size and Weight of the Railroad Compass, Double Vernier.

We make two sizes of this instrument, viz.: five, and five and a half inch needle; the largest size, including the brass head of the jacob staff, weighing eleven pounds, and the five inch, ten and a half pounds.

We invite especial attention to the different styles of our Railroad compasses, believing that in many respects they are very much superior to any other compass made, having a horizontal limb, and an arrangement by which the variation of the needle can be so readily set off and ascertained.

RAILROAD COMPASS.

Single Vernier, 5 inch.

Made by

W. & L. E. GURLEY,

TROY, N.Y.

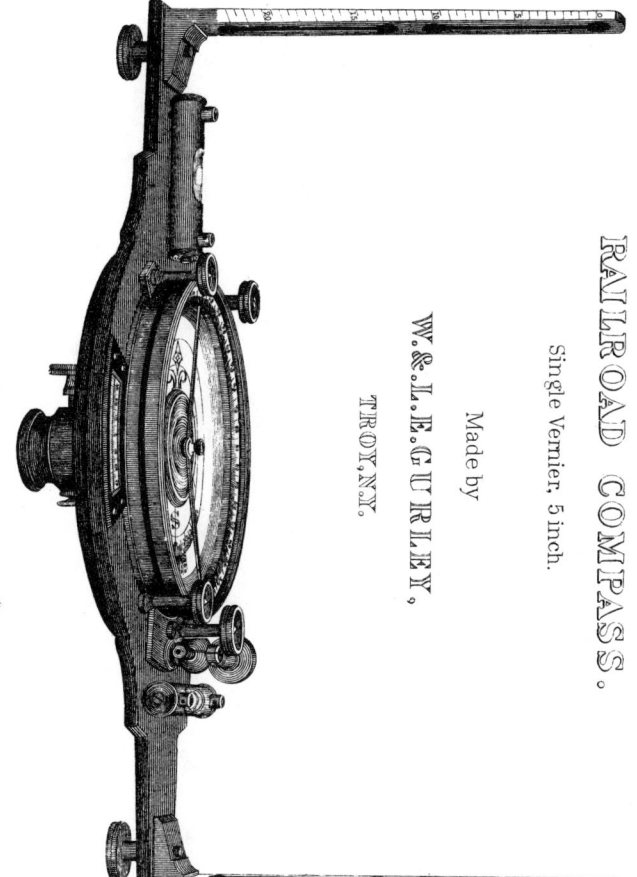

Price as shown above $65.00.

BENJ. D. BENSON, N.Y.

SURVEYOR'S TRANSIT.

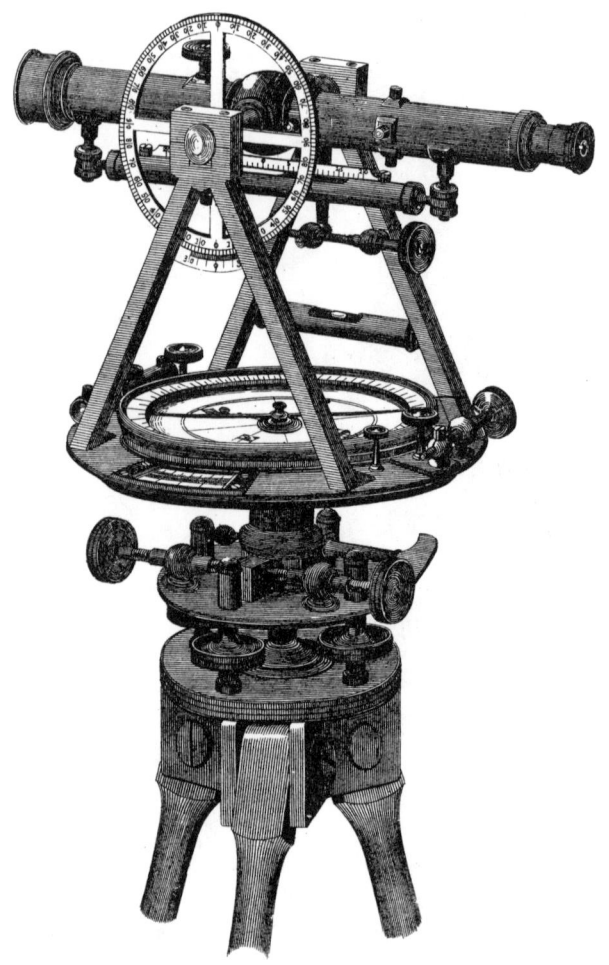

Price as shown above $200,00

Made by

W. & L. E. GURLEY,

TROY, N. Y.

BENJ. D. BENSON, N.Y.

THE SURVEYOR'S TRANSIT.

This instrument shown in the engraving on the opposite page, is in principle very similar to the instrument just described, differing from it mainly in the substitution of the telescope with its appendages, for the ordinary compass sights.

THE TELESCOPE is of somewhat finer quality than that used with the Vernier Transit; as here shown, it is furnished with a small level, having a ground bubble tube and a scale; and also a vertical circle connected with its axis.

THE STANDARDS are made precisely like those of the Vernier Transit, the bearings of the axis of the telescope being conical, and fitted with the utmost nicety; there is also in one of them the moveable piece for the adjustment of the wires to the tracing of a vertical line.

THE SPIRIT LEVELS are placed upon the upper surface of the vernier plate, one being fixed on the standard so as not to obstruct the light which falls on the vernier opening beneath.

Both levels are adjustable with the ordinary steel pin.

THE NEEDLE is like that of the previous instrument, and is five inches long, as shown.

THE VERNIER PLATE, which carries the verniers and telescopes, is made to move with perfect ease and stability, around the graduated circle or limb, and horizontal angles are taken to single minutes; the variation of the needle is also set off by the pinion and clamp screw, as described in the account of the previous instrument.

THE VERNIERS, as in all our angular instruments, are double, reading either way from the centre mark, and to single minutes of a degree.

There are two verniers, placed on opposite sides of the instrument at right angles to the telescope; only one of these is shown in the cut.

THE DIVIDED CIRCLE, or limb, is graduated to half degrees, reads to minutes by the verniers, and is figured as described before.

THE CLAMP AND TANGENT movement of the vernier plate is the same as that of the Railroad Compass; it is well shown in the figure.

THE LEVELING HEAD.—This instrument, as shown in the engraving, is generally used on a tripod.

THE LIGHT LEVELING TRIPOD, used with the Surveyor's Transit, is well shown in the engraving. As there seen, there are nuts screwed in to the upper parallel plate, so as to give a long bearing for the four leveling screws.

The under plate supports the feet of the screws, and has beneath a cavity or bowl, in which moves a hemispherical nut screwed to the spindle of the tripod.

This nut serves both to connect the plates together, and as a pivot on which the upper plate is turned by the leveling screws.

The under parallel plate has also a screw on the under side, by which the tripod head may be disconnected from the legs, and packed in the box with the instrument.

The leveling screws are made of bell metal, have a large double milled head, and a deep screw of about forty threads to the inch; their ends set into little brass cups, so that the screws are worked without indenting the under plate. Sometimes a piece of leather is put in place of the cups.

The leveling screws are entirely covered above by little caps which screw over the upper side of the nut.

When the screws are loosened, the upper plate can be shifted around, so as to bring the leveling screws in any position, with reference to the plates and telescope of the instrument.

The clamp and tangent screws are seen on the upper plate of the tripod. In place of the single tangent screw, we have

in all our instruments, substituted the double tangent movement, as shown in the engraving.

The spindle of the tripod head rises above the upper plate,
and the instrument can be removed from it, by pulling out
a little pin made to spring into a groove, and thus keep the
instrument from falling, when the tripod is carried upon the
shoulder.

In the lower end of the spindle and underneath the plates,
is screwed the loop for attaching the string of the plumb-bob

To LEVEL THE TRIPOD, the engineer takes hold of the oppo
site screw heads with the thumb and fore finger of each hand,
and turning both thumbs in or out, as may be necessary,
raises one side of the upper parallel plate and depresses the
other, until the desired correction is made.

SHIFTING PLATE TO LEVELING TRIPOD HEAD.—In this ar-

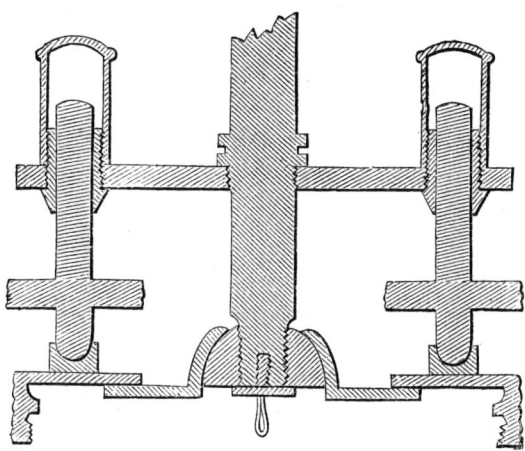

rangement, as shown
in the cut, the lower
leveling-plate is made
up of two separate
pieces, the principal
one screwing on the
plate to which the legs
are connected, and the
smaller, or shifting
piece, in which the
hemispherical nut,
connecting the stem
of the leveling head,
turns — as the plates
are leveled in ordinary
use.

It will be readily seen that when the point of the plummet falls to either
side of a desired point, that by loosening the leveling screws, the little
plate, and the whole instrument with it, can be easily moved in any direction, and thus the plummet brought precisely upon the desired point,
without disturbing the tripod legs. The plate admits of about one inch
movement.

ADJUSTING SOCKET, a beautiful arrangement for occasional

use in place of the leveling tripod, in cases where greatei lightness and rapidity of adjustment are desired, is shown in the adjusting socket, described in the account of the Solai Compass.

To adjust the Surveyor's Transit.

The Levels are adjusted with a steel pin as those of the Vernier Transit, and it need only be added here, that in this, as well as other instruments having two plates, moving upon sockets independent of each other, the levels, when adjusted on one plate, should still keep their position when both are clamped together and turned upon a common socket.

Otherwise, however accurately the telescope might trace a vertical line, when revolved upon the socket of one plate, it would give a very different result as soon as the position of the other plate was changed.

The Needle and telescope with its other attachments being adjusted, as described in our account of the Vernier Transit, we shall here consider only that of the

Level on Telescope.—For the adjustment of this attachment we shall give two methods, the first being that usually practiced by us.

1. First level the instrument carefully, and with the clamp and tangent movement to the axis, make the telescope horizontal as near as may be with the eye, then having the line of collimation previously adjusted, drive a stake at a convenient distance, say from one to three hundred feet, and note the height cut by the horizontal wire, upon a staff set on the top of the stake.

Fix another stake in the opposite direction, and at the same distance from the instrument, and without disturbing the telescope, turn the instrument upon its spindle, set the staff upon the stake, and drive in the ground, until the same height is indicated as in the first observation.

The top of the two stakes will then be in the same heri

zontal line, however much the telescope may be out of level.

Now remove the instrument from fifty to one hundred feet to one side of either of the stakes, and in line with both; again level the instrument, clamp the telescope as nearly horizontal as may be, and note the heights indicated, upon the staff placed first upon the nearest, and then upon the most distant stake.

If both agree, the telescope is level; if not, with the tangent screw move the wire over nearly the whole error, as shown at the distant stake, and repeat the observation as just described. Proceed thus until the horizontal wire will indicate the same height at both stakes, when the telescope will be truly horizontal.

Taking care not to disturb its position, bring the bubble into the centre by the little leveling nuts at the end of the tube, when the adjustment will be completed.

2. Choose a piece of ground nearly level, and having set the instrument firmly, level the plates carefully, and bring the bubble of the telescope into the centre with the tangent screw. Measure in any direction from the instrument, from one to three hundred feet, and drive a stake, and on the stake set a staff and note the height cut by the horizontal wire, then take the same distance from the instrument in an oppo site direction, and drive another stake.

On that stake set the staff and note the height cut by the wire when the telescope is turned in that direction.

The difference of the two observations is evidently the difference of level of the two stakes.

Set the instrument over the lowest stake, or that upon which the greatest height was indicated, and bring the evels on the plates and telescope into adjustment as at first

Then with the staff, measure the perpendicular distance from the top of the stake to the centre cf one of the horizontal

cross wire screw heads; from that distance subtract the difference of level between the two stakes and mark the point
on the staff thus found; place the staff on the other stake,
and with the tangent screw bring the horizontal wire to the
mark just found, and the line will be level.

The telescope now being level, bring the bubble of the
level into the centre, by turning the little nuts at the end of
the tube, and noting again if the wires cut the point on the
staff; screw up the nuts firmly and the adjustment will be
completed.

With such a level carefully adjusted, the engineer, by taking equal fore and back sights, can run horizontal lines with
great rapidity, and a good degree of accuracy.

To use the Surveyor's Transit.

In surveying with this instrument, the plates must be set so
that the zeros of the circle and the verniers correspond, and
firmly clamped together, the eye end of the telescope being
placed over the south side of the compass circle, in the posi
tion shown in the engraving.

The surveyor may then proceed precisely as with the plain
compass.

To turn off Angles.—When angles are to be measured
independently of the needle, proceed precisely as directed in
the description of the Railroad Compass.

The Variation of the Needle is also set off as mentioned in our account of that instrument.

Sizes of the Surveyor's Transit.

We make three sizes of this instrument, the weights and
dimensions of limb of each being as follows:

4 inch needle horizontal limb, 6 in diameter, weight 12½ lbs.
5 "   "   6½ "   " 15 lbs.
5½ "   "   7 "   " 15½ lbs.

(Leveling head included in weight.)

SURVEYOR'S TRANSIT.

Single Vernier with Shifting Tripod Head

Made by

W. & L. E. GURLEY.

TROY, N.Y.

Price as shown above $161.00.

Single Vernier Surveyor's Transit.

We have just introduced a modification of this favorite instrument, by which, with a lighter socket and one double vernier to the limb, we furnish all the capabilities of the more costly instrument, at a material reduction in price.

We make four sizes of this transit, of the same dimensions as those having two verniers to the limb; the engraving opposite represents the one with five inch needle, having also a level attached to telescope, with clamp and tangent to axis.

This instrument may be used on the ball spindle and compass tripod, like the Vernier Transit, but like other transits should also be furnished with the usual leveling tripod.

Sizes and Weights of Single Vernier Surveyor's Transit.

4 inch needle, with leveling head, without tripod, 12 lbs.
4½   "   "   "   13½ lbs.
5   "   "   "   14 lbs.
5½   "   "   "   15 lbs.

The Single Vernier Surveyor's Transit, from its lightness, excellence, and cheapness, has well supplied a need long felt by engineers and surveyors, in furnishing an instrument suitable for accurate work at a very reasonable cost.

Merits of the Surveyor's Transit.

In this instrument, as just described, the surveyor will recognize advantages not possessed by any other instrument with which we are acquainted.

Combining the capabilities of a needle instrument, with a fine telescope, and the accuracy of a divided limb and verniers, and having also the means for turning off the variation of the needle; it is for a mixed practice of accurate surveying and engineering, such indeed as is required by most city engineers, the best instrument ever constructed.

The peculiar construction of the sockets and plates of this instrument is entirely our own invention, and we feel the utmost confidence in recommending it to all whose practice is such as to require the use of the needle combined with that of the divided circle and verniers.

70 THE SOLAR COMPASS.

THE SOLAR COMPASS.

This instrument, so ingeniously contrived for readily deter mining a true meridian or north and south line, was invented by WILLIAM A. BURT, of Michigan, and patented by him in 1836.

It has since come into general use in the surveys of U. S. public lands, the principal lines of which are required to be run with reference to the true meridian.

The invention having long since become the property of the public, we have given our attention to the manufacture of these instruments, and are now prepared to furnish them, with important improvements of our own devising, at greatly reduced prices.

Our improved Solar Compass, one form of which is shown in the engraving, has nearly the same arrangement of plates, with divided circles, verniers, and sockets, as the Railroad Compass.

The Solar Apparatus.

The Solar Apparatus is seen in the place of the needle, and in fact operates as its substitute in the field.

It consists mainly of three arcs of circles, by which can be set off the latitude of a place, the declination of the sun, and the hour of the day.

These arcs, designated in the cut by the letters a, b, and c, are therefore termed the latitude, the declination, and the hour arcs respectively.

THE LATITUDE ARC, a, has its centre of motion in two pivots one of which is seen at d, the other is concealed in the cut.

It is moved either up or down within a hollow arc, seen in the cut, by a tangent screw at f, and is securely fastened in any position by a clamp screw.

The Latitude arc is graduated to quarter degrees, and reads by its vernier, e, to single minutes ; it has a range of about thirty-five degrees, so as to be adjustable to the latitude of any place in the United States.

SOLAR COMPASS,

Made by

W. & L. E. GURLEY,

TROY, N. Y.

BENJ. O. BENSON. N.Y.

Price of the above, including tripod with parallel plates and leveling screws $220.00.

The Declination Arc, *b*, is also graduated to quarter degrees, and has a range of about twenty-four degrees.

Its vernier, *v*, reading to single minutes, is fixed to a movable arm, *h*, having its centre of motion in the centre of the declination arc at *g* ; the arm is moved over the surface of the declination arc, and its vernier set to any reading by turning the head of the tangent screw, *k*. It is also securely clamped in any position by a screw, concealed in the engraving.

Solar Lenses and Lines.—At each end of the arm, *h*, is a rectangular block of brass, in which is set a small convex lens, having its focus on the surface of a little silver plate, fastened by screws to the inside of the opposite block.

The silver plate, with its peculiar lines. will be referred to more particularly hereafter.

Equatorial Sights.—On the top of each of the rectangular blocks is seen a little sighting piece, termed the equatorial sight fastened to the block by a small milled head screw, so as to be detached at pleasure.

They are used, as will be explained hereafter, in adjusting the different parts of the solar apparatus.

The Hour Arc, *c*, is supported by the two pivots of the latitude arc, already spoken of, and is also connected with that arc by a curved arm, as shown in the figure.

The hour arc has a range of about 120°, is divided to half degrees, and figured in two series ; designating both the hours and the degrees, the middle division being marked 12 and 90 on either side of the graduated lines.

The Polar Axis.—Through the centre of the hour arc passes a hollow socket, *p*, containing the spindle of the declination arc, by means of which this arc can be moved from side to side over the surface of the hour arc, or turned completely round as may be required.

The hour arc is read by the lower edge of the graduated side of the declination arc

The axis of the declination arc, or indeed the whole socket *p*, is appropriately termed the polar axis.

THE ADJUSTER.—Besides the parts shown in the cut, there is also an arm used in the adjustment of the instrument as described hereafter, but laid aside in the box when that is effected.

The parts just described constitute properly the solar apparatus.

Besides these, however, are seen the needle box, *n*, with its arc and tangent screw, *t*, and the spirit levels, for bringing the whole instrument to a horizontal position.

THE NEEDLE BOX, *n*, has an arc of about 36° in extent, divided to half degrees, and figured from the centre or zero mark on either side.

The needle, which is made as in other instruments, except that the arms are of unequal lengths, is raised or lowered by a lever shown in the cut.

The needle box is attached by a projecting arm to a tangent screw, *t*, by which it is moved about its centre, and its needle set to any variation.

This variation is also read off by the vernier on the end of the projecting arm, reading to single minutes a graduated arc, attached to the plate of the compass.

THE LEVELS seen with the solar apparatus, have ground glass vials, and are adjustable at their ends like those of our other instruments.

The edge of the circular plate on which the solar work is placed, is divided and figured at intervals of ten degrees, and numbered, as shown, from 0 to 90 on each side of the line of sight.

These graduations are used in connection with a little brass pin, seen in the centre of the plate, to obtain approximate bearings of lines, which are not important enough to require a close observation.

LINES OF REFRACTION.—The inside faces of the sights are also graduated and figured, to indicate the amount of refraction to be allowed when the sun is near the horizon. These are not shown in the cut.

THE HORIZONTAL LIMB in all our Solar Compasses is divided upon silver, and reads by two opposite verniers to single minutes of a degree, the number of minutes being counted off in the same direction in which the vernier moves.

Definition of Astronomical Terms.

Before proceeding to describe the principles and adjustments of this instrument, a brief statement of the terms employed may here be appropriately made.

THE SUN is the centre of the solar system, remaining constantly fixed in its position, though, for the sake of convenience, often spoken of as in motion around the earth.

THE EARTH makes a complete revolution around the sun in 365 days, 6 hours, very nearly.

It also rotates about an imaginary line passing through its centre, and termed its *axis*, once in twenty-four hours, turning from west to east.

THE POLES are the extremities of the axis ; that in our own hemisphere, known as the north pole, if produced indefinitely towards the concave surface of the heavens, would reach a point situated near the polar star, and called the north pole of the heavens.

THE EQUATOR is an imaginary line passing around the earth equi-distant from the poles, and at right angles with them.

If the plane of the equator is produced to the heavens, it forms what is termed the equator of the heavens.

THE ORBIT of the earth is the path in which it moves in making its yearly revolution.

If the plane of this orbit were extended to the heavens, it would form the *ecliptic* or the sun's apparent path in the heavens.

The earth's axis is inclined to its orbit at an angle of about 23° 28', making the angle between the earth's orbit and its equator, or between the celestial equator and the ecliptic of the same amount.

THE EQUINOXES are the two points in which the ecliptic and the celestial equator intersect each other.

THE DECLINATION of the sun is its angular distance north or south of the celestial equator ; when the sun is at the equinoxes, that is about the 21st of March and the 21st of September of each year, his declination is 0, or he is said to be on the equator; from these points his declination increases from day to day, and from hour to hour, until, on the 21st of June and 21st of December, he is 23° 28' distant from the equator.

It is the declination which causes the sun to appear so much higher in summer than in winter, his altitude in the heavens being in fact nearly 47° more on the 21st of June than it is on the 21st of December.

THE HORIZON of a place is the surface which is defined by a plane supposed to pass through the place at right angles to a vertical or plumb line, and to bound our vision at the surface of the earth.

The horizon or a horizontal surface is determined by the surface of any liquid when at rest, or by the spirit levels of an instrument.

THE ZENITH of any place is the point directly over head, at right angles to the horizon.

THE MERIDIAN of any place is a great circle passing through the zenith of a place, and the poles of the earth.

The meridian, or true north and south line of any place is the line determined by the intersection of the plane of the meridian circle with the plane of the horizon.

THE MERIDIAN ALTITUDE of the sun is its angular elevation above the horizon, when passing the meridian of a place

THE LATITUDE of a place is its distance north or south of the equator, measured on a meridian. At the equator the latitude is 0°, at the poles 90°.

THE LONGITUDE of a place is its distance in degrees or in time, east or west of a given place taken as the starting point or first meridian; it is measured on the equator or any parallel of latitude.

In the Nautical Almanac, which is commonly used with the Solar Compass, the longitude of the principal places in the United States is reckoned from Greenwich, England, and expressed both in degrees and hours.

THE ZENITH DISTANCE of any heavenly body, is its angular distance north or south of the zenith of a place, measured when the body is on the meridian.

Suppose a person situated on the equator at the time of the equinoxes, the sun, when on the meridian, would be in the zenith of the place, and the poles of the earth would, of course, lie in the plane of his horizon.

Disregarding for the present the declination of the sun, let us suppose the person travels towards the north pole.

As he passes to the north, the sun will descend from the zenith, and the pole rise from the horizon in the same proportion, until when he arrives at the north pole of the earth, the sun will have declined to the horizon, and the pole of the heavens will have reached the zenith.

The altitude of the pole at any place, or the distance of the sun from the zenith, would in the case supposed, give the observer the latitude of that place.

If we now take into account the sun's declination, it would increase or diminish its meridian altitude, according as it passes north or south of the equator ; but the declination of the sun at any time being known, its zenith distance, and therefore the latitude of the place, can be readily ascertained by an observation made when it is on the meridian.

As we shall see hereafter, it is by this method that we obtain the latitude of any place by the Solar Compass.

TIME.—A solar day is the interval of time between the departure of the sun from the meridian of a place, and its succeeding return to the same position.

The length of the solar day, by reason of the varying velocities of the earth in its orbit, and the inclination of its axis, is continually changing.

In order to have a uniform measure of time, we have recourse to what is termed a *mean solar day*, the length of which is equal to the mean or average of all the solar days in a year.

The time thus given is termed *mean time*, and is that to which clocks and watches are adjusted for the ordinary business of life.

The sun is sometimes faster, and sometimes slower, than the clock, the difference being termed the *equation of time.*

The moment when the sun is on the meridian of any place is termed *apparent noon,* and this being ascertained, we can, by referring to the equation of time for the given day, and adding to, or subtracting from, apparent noon, according as the sun is slow or fast, obtain the time of *mean noon,* by which to set the watch or chronometer.

DIFFERENCE OF LONGITUDE.—As the earth makes a complete rotation upon its axis once a day, every point on its surface must past over 360° in 24 hours, or 15° in one hour, and so on in the same proportion.

And as the rotation is from west to east, the sun would come to the meridian of every place 15° west of Greenwich, just one hour later than the time given in the Almanac, for apparent noon at that place.

To an observer situated at Troy, N. Y., the longitude of which is in time 4 hours 54 minutes, 40 sec., the sun would come to the meridian nearly five hours later than at Greenwich, and thus when it was 12 M. at that place, it would be be but about 7 o'clock A. M. in Troy.

REFRACTION.—By reason of the increasing density of the atmosphere from its upper regions to the earth's surface, the rays of light from the sun are bent out of their course, so as to make his altitude appear greater than is actually the case.

The amount of refraction varies, according to the altitude of the body observed ; being 0 when it is in the zenith, about one minute when midway from the horizon to the zenith, and almost 34' when in the horizon.

To indicate the amount of refraction to be allowed in observations with the solar compass, the sights have on their inside faces a number of lines, marked at intervals, and figured so as to read off the degree of refraction of the sun or other body, when seen directly over the top of one sight, by the eye placed on the other at the height marked by the line of refraction.

EFFECT OF INCIDENTAL REFRACTION.—It will be seen by referring to the instrument, that the effect of the ordinary refraction, which to distinguish· from meridional refraction, we will term *incidental*, upon the position of the sun's image with reference to the equatorial lines, which, in fact, are the only ones to be regarded in running lines with the Solar Compass, is continually changing, not only with the change of latitude, but also with that of the sun's declination from hour to hour, and the motion of the revolving arm as it follows the sun in its daily revolution.

If the equatorial lines were always in the same vertical plane with the sun, as would be the case at the equator at the time of the equinoxes, it is evident that refraction would have no effect upon the position of the image between these lines, and therefore would not be of any importance to the surveyor.

But as we proceed further north, and as the sun's declination to the south increases, the refraction also increases, and must now be taken into account.

Again the angle which the equatorial lines make with the horizon is continually changing as the arm is made to follow the motion of the sun during the course of a day.

Thus in the morning and evening they are more or less inclined to the horizon, while at noon they are exactly parallel to it.

And thus it follows that the excess of refraction at morning and evening is in some measure balanced by the fact that the position of the sun's image with reference to the equatorial lines is then less affected by it, on account of the greater inclination of the lines to the horizon.

Besides the causes already mentioned as modifying the effect of refraction, we may add that of the numerous changes in the atmosphere ; and when all these are considered, it will be seen that any idea of calculating the amount and influence of refraction with certainty may well be abandoned.

The best than can be done, is to estimate and allow for it as a little experience will suggest during morning and evening, and disregard it entirely for the rest of the day.

We shall give the practice of government surveyors in allowing for refraction, when we come to speak of the manner in which the Solar Compass is used.

THE MERIDIONAL REFRACTION of the sun is that by which his position is affected when on the meridian.

It varies of course, with the varying meridional altitudes of the sun, and in all places in north latitude must be added to the declination when it is north, and subtracted when it is south, in order to obtain the true declination for the given day.

This is done for the starting hour as will be hereafter described, and is carried through the other hours of the day without further addition.

To obtain the meridional refraction of the sun for a given

day and place we first ascertain the meridional altitude of
the sun on that day, and then refer to a table of refractions in Burt's Key to the Solar Compass, or Clevenger's
Government Surveying,* to find the amount of refraction
corresponding to the angle thus obtained, which will be
the meridional refraction required.

Thus to obtain the meridional refraction of the sun for
April 16th, 1874, at Troy, N. Y., proceed as follows:

Latitude of Troy..... .....42° 44′

90—42° 44′=17° 16′

Declination April 16, which is north or+....10° 9′ 46″

Gives a meridional altitude of 57° 6′ 14″

Referring now to the table of refractions we find the
amount corresponding to the angle thus found to be 38″,
which is the meridional refraction for the given day, and is
to be added to the declination, as will hereafter be seen.

Again, let it be required to obtain the meridional refraction at the same place, Oct. 16th, 1874, the declination now
being in the contrary direction.

Latitude of Troy.......... 42° 44′

90—42° 44′=47° 16′

Declination for Oct. 16, which is south or—.... 8° 55′ 50″

Gives a meridional altitude of 38° 20′ 10′

Referring again to the table, we find the refraction for the
meridional altitude just found to be 1′ 14″, which is now to
be subtracted from the declination for the given day, when
it is set off upon the declination arc.

As will be seen by both examples, the meridional refraction,
though affecting the sun's declination by a constant quantity
through the whole day, is yet of comparatively small amount,
and in practice is often entirely disregarded, except by those
surveyors engaged in running the great standard meridians
and parallels of latitude, or in testing their instruments

* *See* also p. 102 of this Manual

Principles of the Solar Compass.

We are now prepared, to proceed to the explanation of the peculiar construction of the instrument we are considering.

The little silver plate before referred to, is shown in detail in fig. 13. On its surface are marked two sets of lines inter-

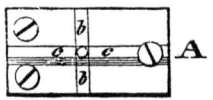

Fig. 13.

secting each other at right angles; of these, *b b* are termed the hour lines, and *c c* the equatorial lines, as having reference respectively, to the hour of the day and the position of the sun, in relation to the equator.

Below the equatorial lines are also marked three other lines, which are five minutes apart, and are of service in making allowance for refraction, as will be hereafter explained.

The interval between the two lines *c, c,* as well as between *b, b,* is just sufficient to include the circular image of the sun, as formed by the solar lens, on the opposite end of the revolving arm.

When, therefore, the instrument is made perfectly horizontal, the equatorial lines and the opposite lenses being accurately adjusted to each other by a previous operation, and the sun's image brought within the equatorial lines, his position in the heavens, with reference to the horizon, wil' be defined with precision

Suppose the observation to be made at the time of one of the equinoxes ; the arm *h,* set at zero on the declination arc *b,* and the polar axis *p,* placed exactly parallel to the axis the earth.

Then the motion of the arm *h,* if revolved on the spindle of the declination arc around the hour circle *c,* will exactly correspond with the motion of the sun in the heavens, on the given day and at the place of observation ; so that if the sun's image was brought between the lines *c c,* in the morning, it would continue in the same position, passing neither above nor below the lines, as the arm was made to revolve in imitation of the motion of the sun about the earth.

In the morning as the sun rises from the horizon, the arm *h* will be in a position nearly at right angles to that shown in the cut, the lens being turned towards the sun, and the silver plate on which his image is thrown directly opposite.

As the sun ascends, the arm must be moved around, until when he has reached the meridian, the graduated side of the declination arc will indicate 12 on the hour circle, and the arm *h*, the declination arc *b*, and the latitude arc *a*, will be in the same plane.

As the sun declines from the meridian the arm *h* must be moved in the same direction, until at sunset its position will be the exact reverse of that it occupied in the morning.

ALLOWANCE FOR DECLINATION.—Let us now suppose the observation made when the sun has passed the equinoctial point, and when his position is affected by declination.

By referring to the Almanac, and setting off on the arc his declination for the given day and hour, we are still able to determine his position with the same certainty as if he remained on the equator.

When the sun's declination is south, that is from the 22d of September to the 20th of March in each year, the arc *b* is turned towards the plates of the compass, as shown in the engraving, and the solar lens, *o*, with the silver plate opposite, are made use of in the surveys.

The remainder of the year, the arc is turned from the plates, and the other lens and plate employed.

When the Solar Compass is accurately adjusted, and its plates made perfectly horizontal, the latitude of the place, and the declination of the sun for the given day and hour, being also set off on the respective arcs, *the image of the sun cannot be brought between the equatorial lines until the polar axis is placed in the plane of the meridian of the place, or in a portion parallel to the axis of the earth.* The slightest deviation from this position will cause the image to pass above or below the lines, and thus discover the error.

We thus, from the position of the sun in the solar system, obtain a certain direction absolutely unchangeable, from which to run our lines, and measure the horizontal angles required.

This simple principle is not only the basis of the construction of the Solar Compass, but the sole cause of its superiority to the ordinary or magnetic instrument. For in a needle instrument, as before stated in this work, the accuracy of the horizontal angles indicated, and therefore of all the observations made, depends upon " the delicacy of the needle, and the constancy with which it assumes a certain direction, termed the magnetic meridian."

The principal causes of error in the needle as briefly stated, are the dulling of the pivot, the loss of polarity in the needle, the influence of local attraction, and the effect of the sun's rays, producing the diurnal variation.

From all these imperfections the solar instrument is free.

The sights and the graduated limb being adjusted to the solar apparatus, and the latitude of the place, and the declination of the sun also set off upon the respective arcs, we are able, not only to run the true meridian, or a due east and west course, but also to set off the horizontal angles with minuteness and accuracy from a direction which never changes, and is unaffected by attraction of any kind.

To adjust the Solar Compass.

The adjustments of this instrument, with which the surveyor will have to do, are simple and few in number, and will now be given in order.

1st, To ADJUST THE LEVELS.—Proceed precisely as directed in the account of the other instruments we have described, by bringing the bubbles into the centre of the tubes by the leveling screws of the tripod, and then reversing the instrument upon its spindle, and raising or lowering the ends of the tubes, until the bubbles will remain in the centre during a complete revolution of the instrument.

2d, To ADJUST THE EQUATORIAL LINES AND SOLAR LENSES.— First detach the arm *h* from the declination arc by withdrawing the screws shown in the cut from the ends of the posts of the tangent screw, *k*, and also the clamp screw, and the conical pivot with its small screws by which the arm and declination arc are connected.

The arm, *h*, being thus removed, attach the adjuster in its place by replacing the conical pivot and screws, and insert the clamp screw so as to clamp the adjuster at any point on the declination arc.

Now level the instrument, place the arm *h* on the adjuster, with the same side resting against the surface of the declination arc as before it was detached. Turn the instrument on its spindle so as to bring the solar lens to be adjusted in the direction of the sun, and raise or lower the adjuster on the declination arc, until it can be clamped in such a position, as to bring the sun's image as near as may be between the equatorial lines on the opposite silver plate, and bring the image precisely into position, by the tangent of the latitude arc, or the leveling screws of the tripod. Then carefully turn the arm half way over, until it rests upon the adjuster by the opposite faces of the rectangular blocks, and again observe the position of the sun's image.

If it remains between the lines as before, the lens and plate are in adjustment; if not, loosen the three screws which confine the plate to the block, and move the plate under their heads, until one half the error in the position of the sun's image is removed.

Again bring the image between the lines, and repeat the operation until it will remain in the same situation, in both positions of the arm, when the adjustment will be completed

To adjust the other lens and plate, reverse the arm end for end on the adjuster, and proceed precisely as in the former case, until the same result is attained.

In tightening the screws over the silver plate, care must be taken not to move the plate.

This adjustment now being complete, the adjuster should be removed, and the arm, h, with its attachments, replaced as before.

3d, To Adjust the Vernier of the Declination Arc.— Having leveled the instrument, and turned its lens in the direction of the sun, clamp to the spindle, and set the vernier, v, of the declination arc, at zero, by means of the tangent screw, at k, and clamp to the arc.

See that the spindle moves easily and yet truly in the socket, or polar axis, and raise or lower the latitude arc by turning the tangent screw, f, until the sun's image is brought between the equatorial lines on one of the plates. Clamp the latitude arc by the screw, and bring the image precisely into position by the leveling screws of the tripod or socket, and without disturbing the instrument, carefully revolve the arm h, until the opposite lens and plate are brought in the direction of the sun, and note if the sun's image comes between the lines as before.

If it does, there is no index error of the declination arc ; if not with the tangent screw, k, move the arm until the sun's image passes over half the error; again bring the image between the lines, and repeat the operation as before, until the image will occupy the same position on both the plates.

We shall now find, however, that the zero marks on the arc and the vernier do not correspond, and to remedy this error, the little flat head screws above the vernier must be loosened until it can be moved so as to make the zeros coincide, when the operation will be completed.

4th, To Adjust the Solar Apparatus to the Compass Sights.—First level the instrument, and with the clamp and tangent screws set the main plate at 90° by the verniers and horizontal limb. Then remove the clamp screw, and raise the latitude arc until the polar axis is by estimation very nearly horizontal, and if necessary, tighten the screws on the pivots of the arc, so as to retain it in this position.

Fix the vernier of the declination arc at zero, and direct the equatorial sights to some distant and well marked object, and observe the same through the compass sights. If the same object is seen through both, and the verniers read to 90° on the limb, the adjustment is complete ; if not, the correction must be made by moving the sights or changing the position of the verniers.

In that form of this instrument, described hereafter as the Solar Compass proper, the solar work is attached permanently to the sockets, and this adjustment once made by the maker, will need no further attention at the hands of the surveyor, unless in case of severe accidents.

Other Adjustments.—We should perhaps here say, that the above adjustments, as well as the others with which the surveyor ordinarily will have no concern, are all made by us, in the process of constructing and finishing our instruments, and are liable to very little derangement in the ordinary use of the Solar Compass.

Tripods, &c.

The Solar Compass should always be used on a tripod provided with some means by which it may be leveled with ease and accuracy.

A tangent motion to the whole instrument about its spindle.

in addition to that of the limb already spoken of, is also of very great value.

These requirements are, we think, best supplied in our Adjusting Socket, with compound tangent ball, shown in fig. 14, being screwed into the top of a tripod like the ordinary leveling head.

The interior stem of the socket is expanded above to receive the ball of the compass, and below, pivots upon a small ball confined underneath the plate of the tripod.

The instrument is approximately leveled by the ball and socket joint, and finally made perfectly horizontal by the leveling screws of the socket.

Fig. 14

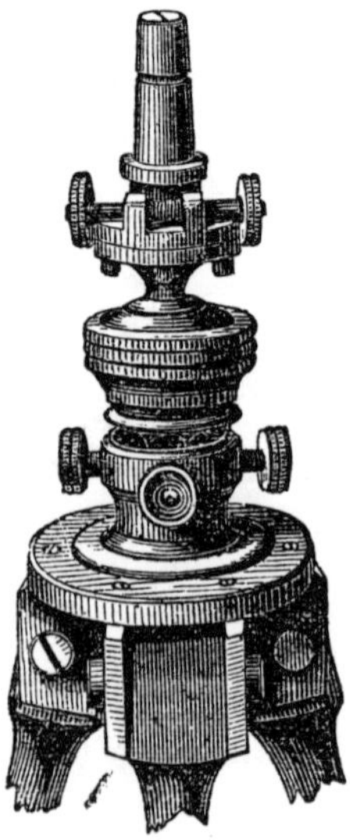

It also revolves upon the spindle as upon the ordinary compass ball, but can be clamped at pleasure to the spindle, and then by its sights or telescope directed precisely to any object by the tangent screw of the compound ball.

The ordinary adjusting tripod head with leveling screws, and clamp and tangent movement, as shown with our Surveyor's Transit, is also used with this instrument, but is heavier and less capable of rapid adjustment.

Of course when a single or jacob staff is preferred to the tripod, for this or any other of our compasses, the adjusting socket, with either a simple or compound ball, can be placed upon the top of the staff and adjusted as just described.

In cases where either a leveling tripod or the adjusting socket, with compound ball, is furnished with this instrument, a simple ball is also supplied, upon which the compass can be placed when desired.

The simple ball is furnished with an extra cap for the socket, or in case of the leveling tripod, with an adopter, fitting to the top of the tripod, so that the substitution can be made by the surveyor himself without any difficulty.

PRICES OF THE SOCKET.—As the adjusting socket may often be used to advantage with the other instruments described in this work, we will here insert its prices with its various modifications.

With tripod and compound ball as in fig. 14 ..$18 00
  " " simple " 12 00
 " Jacob staff and compound ball......... 12 00
  "   "   simple " 6 00

To use the Solar Compass.

Before this instrument can be used at any given place it is necessary to set off upon its arcs both the declination of the sun as affected by its meridional refraction for the given day, and the latitude of the place where the observation is made.

To SET OFF THE DECLINATION.—The declination of the sun, given in the Ephemeris of the Nautical Almanac, from year to year, is calculated for apparent noon at Greenwich, England.

To determine it for any other hour at a place in the U. S., reference must be had, not only to the difference of time arising from the longitude, but also to the change of declination from day to day.

The longitude of the place, and therefore its difference in time, if not given directly in the tables of the Almanac, can be ascertained very nearly by reference to that of other places given, which are situated on, or very nearly on, the same meridian.

It is the practice of surveyors in the States east of the Mississippi to allow a difference of *six* hours for the difference in longitude, calling the declination given in the Almanac for 12 M., that of 6 A. M., at the place of observation.

Beyond the parallel of Santa Fe, the allowance would be about *seven* hours, and in California, Oregon, and Washington Territory about *eight* hours

Having thus the difference of time, we very readily obtain the declination for a certain hour in the morning, which would be earlier or later as the longitude was greater or less, and the same as that of apparent noon at Greenwich on the given day. Thus suppose the observation made at a place, say, five hours later than Greenwich, then the declination given in the Almanac for the given day at noon, affected by the meridional refraction, would be the declination at the place of observation for 7 o'clock, A. M.; this gives us the starting point.

To obtain the declination for the other hours of the day take from the Almanac, the declination for apparent noon of the given day, and also that of the day following, subtract one from the other, as it may have increased or decreased, and we have the change of declination for 24 hours, divide this by 24, and we obtain the change of declination for a single hour, which is to be added to, or subtracted from that of the starting hour, according as the declination is increasing or decreasing between the two days taken.

To make this more plain we will give an example. Suppose it was required to obtain the declination for the different hours of April 16th, 1874 at Troy, N. Y.

The longitude in time is 4 hrs. 54 min. 40 sec., or **practically 5 hours**, so that the declination given in the Almanac for the given day at Greenwich would be that of 7 A. M. at Troy.

To obtain the hourly change

Say declination at Greenwich, April 17..10° 30′ 57″
"   "   " 16..10 9 46

Change for 24 hours...... 21 11

Reduce to seconds, and divide by 24, and we have an hourly change of 53 seconds, which, as the declination is increasing, is to be added every hour after 7 A. M.

Hence, sun's declination at Greenwich noon as by the
table being..........10° 9′ 46″
Add meridional refraction.   38

			10	10	24	= Dec. for 7 A.M.
Add hourly change.....					53	
			10	11	17	" 8 "
"	"	......			53	
			10	12	10	" 9 "
"	"	......			53	
			10	13	3	" 10 "
"	"	......			53	
			10	13	56	" 11 "
"	"	......			53	
			10	14	49	" 12 M.
"	"	......			53	
			10	15	42	" 1 P. M.
"	"	......			53	
			10	16	35	" 2 "
"	"	......			53	
			10	17	28	" 3 "
"	"				53	
			10	18	21	" 4 "
"	"				53	
			10	19	14	" 5 "

                                  10   19   14=Dec. for 5 P. M.
Add hourly change.....   53
                                  _____
                                  10   20   7   "   .6   "
      "   "   53
                                  _____
                                  10 '21   "   7   "

In the case taken the declination is increasing from day to day, and therefore the hourly change is added; if, on the contrary, the declination was decreasing, the hourly change should be subtracted.

The calculation of the declination for the different hours of the day, should of course be made and noted before the surveyor commences his work, that he may lay off the change from hour to hour, from a table prepared as above described.

It is considered sufficiently accurate by most government surveyors, to set off the declination only three or four times in the day, at intervals of two or three hours as required.

To SET OFF THE LATITUDE.—Find the declination of the sun for the given day at noon, at the place of observation as just described, and with the tangent screw set it off upon the declination arc, and clamp the arm firmly to the arc.

Observe in the almanac the equation of time for the given day, in order to know about the time the sun will reach the meridian.

Then, about fifteen or twenty minutes before this time, set up the instrument, level it carefully, fix the divided surface of the declination arc at 12 on the hour circle, and turn the instrument upon its spindle until the solar lens is brought into the direction of the sun.

Loosen the clamp screw of the latitude arc, and with the tangent screw raise or lower this arc until the image of the sun is brought precisely between the equatorial lines, and turn the instrument from time to time so as to keep the image also between the hour lines on the plate.

As the sun ascends, its image will move below the lines, and the arc must be moved to follow it. Continue thus keeping it between the two sets of lines until its image begins to pass above the equatorial lines, which is also the moment of its passing the meridian.

Now read off the vernier of the arc, and we have the latitude of the place, which is always to be set off on the arc when the compass is used at the given place.

It is the practice of surveyors using the Solar Compass to set off, in the manner just described, the latitude of the point where the survey begins, and to repeat the observation and correction of the latitude arc every day when the weather is favorable, there being also nearly an hour at mid-day when the sun is so near the meridian as not to give the direction of lines with the certainty required.

To Run Lines with the Solar Compass.—Having set off in the manner just given the latitude and declination upon their respective arcs, the instrument being also in adjustment, the surveyor is ready to run lines by the sun.

To do this, the instrument is set over the station and carefully leveled, the plates clamped at zero on the horizontal limb, and the sights directed north and south, the direction being given, when unknown, approximately by the needle.

The solar lens is then turned to the sun, and with one hand on the instrument, and the other on the revolving arm, both are moved from side to side, until the sun's image is made to appear on the silver plate ; when by carefully continuing the operation, it may be brought precisely between the equatorial lines.

Allowance being now made for refraction, the line of sights will indicate the true meridian; the observation may now be made, and the flag-man put in position.

When a due east and west line is to be run, the verniers

of the horizontal limb are set at 90°, and the sun's image kept between the lines as before.

The Solar Compass being so constructed that when the sun's image is in position the limb must be clamped at 0 in order to run a true meridian line, it will be evident that the bearing of any line from the meridian, may be read by the verniers of the limb precisely as in the ordinary magnetic compass, the bearing of lines are read from the ends of the needle.

ALLOWANCE FOR REFRACTION.—From what has been before stated, it will be seen that no precise calculation can be made for the effect of incidental refraction.

The practice of the Government surveyor in this matter is, to keep the image square between the equatorial lines during most of the day, but at morning and evening, when the sun is near the horizon, to run the image full and flush upon the lower line, and pay no regard to the other.

When the sun is near the horizon, the image is less bright and clearly defined than during the rest of the day, and in keeping the brightest part fully on the lower equatorial line, as we have said, the hazy edge of the image will overlap one or two of the graduated spaces below, and thus fully compensate for the effect of refraction.

A little practice with the instrument will soon enable the inexperienced surveyor to supply the correction thus approximately given, so as to make the proper allowance for incidental refraction with more accuracy, than if a precise calculation had been attempted.

USE OF THE NEEDLE.—In running lines, the magnetic needle is alway kept with the sun ; that is, the point of the needle is made to indicate 0 on the arc of the compass box, by turning the tangent screw connected with its arm on the opposite side of the plate. By this means the lines can be run by the needle alone in case of the temporary disappear ance of the sun ; but, of course, in such cases the surveyor must be sure that no local attraction is exerted.

The variation of the needle, which is noted at every station, is read off in degrees and minutes on the arc, by the edge of which the vernier of the needle box moves.

ALLOWANCE FOR THE EARTH'S CURVATURE.—When long lines are run by the Solar Compass, either by the true meridian, or due east and west, allowance must be made for the curvature of the earth.

Thus, in running north or south, the latitude changes about one minute for every distance of 92 chains, 30 links, and the side of a township requires a change on the latitude arc of 5′ 12″, the township, of course, being six miles square.

This allowance is of constant use where the surveyor fails to get an observation on the sun at noon, and is a very close approximation to the truth.

In running due east and west, as in tracing the standard parallels of latitude, the sights are set at 90° on the limb, and the line is run at right angles to the meridian.

If no allowance were made for the earth's curvature, these lines would, if sufficiently produced, reach the equator, to which they are constantly tending.

Of course, in running short lines either east or west, the variation from the parallel, would be so small as to be of no practical importance, but when long sights are taken, the correction should be made by taking fore and back sights at every station, noticing the error on the back sight, and setting off one half of it on the fore sight on the side towards he pole.

TIME OF DAY BY THE SUN.—The time of day is best ascertained by the Solar Compass when the sun is on the meridian, as at the time of making the observation for latitude.

The time thus given is that of apparent noon, and can be reduced to mean time, by merely applying the equation of time as directed in the Almanac, and adding or subtracting as the sun is slow or fast..

The time, of course, can also be taken before or after noon, by bringing the sun's image between the hour lines, and noticing the position of the divided edge of the revolving arm, with reference to the graduations of the hour circle, allowing four minutes of time for each degree of the arc, and thus obtaining apparent time, which must be corrected by the equation of time as just described.

CAUTION AS TO THE FALSE IMAGE.—In using the compass upon the sun, if the revolving arm be turned a little one side of its proper position, a false or reflected image of the sun will appear on the silver plate in nearly the same place as that occupied by the true one. It is caused by the reflection of the true image from the surface of the arm, and is a fruitful source of error to the inexperienced surveyor. It can, however, be readily distinguished from the real image by being much less bright, and not so clearly defined.

APPROXIMATE BEARINGS.—When the bearings of lines, such as the course of a stream, or the boundaries of a forest, are not desired with the certainty given by the verniers and horizontal limb, a rough approximation of the angle they make with the true meridian, is obtained by the divisions on the outside of the circular plate.

In this operation, a pencil, or thin straight edge of any sort, is held perpendicularly against the circular edge of the plate, and moved around until it is in range with the eye, the brass centre pin, and the object observed.

The bearing of the line is then read off at the point where the pencil is placed.

Time for using the Solar Compass.

The Solar Compass, like the ordinary instrument, can be used at all seasons of the year, the most favorable time being, of course, in the summer, when the declination is north, and the days are long, and more generally fair.

It is best not to take the sun at morning and evening, when it is within half-an-hour of the horizon, nor at noon, as we have before stated, for about the same interval, before and after it passes the meridian.

Telescope and Micrometer.

It is often desirable to use a telescope in connection with the Solar Compass, both for ranging lines and measuring distances, where the chain cannot be conveniently employed

Fig. 15.

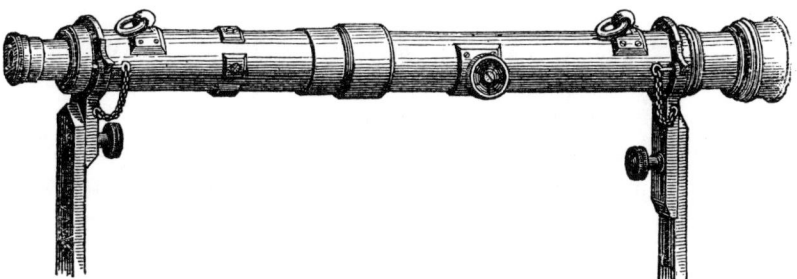

The above cut will show the arrangement which we have commonly made for this purpose, and which is intended to use with that form of the instrument provided with the ordinary sights.

The telescope is from 17 to 20 inches in length, and furnished with the same glasses as those of the best leveling instruments.

It has, of course, the ordinary cross-wires, with screws for their adjustment ; the centering ring for the eye-piece, the screws and washers of which are shown at the eye end; and a pinion for bringing the object glass into focus upon near and distant objects.

The tube of the telescope is placed in Y pieces like those of the leveling instrument, so that it can be revolved at pleasure on loosening the pins which confine it, or entirely removed from the wyes, and carried on the person of the surveyor, by a cord suspended from the little rings shown on its upper surface.

The wye piece on the forward sight has a slote as shown, so as to allow the telescope to move up or down a short distance, and to be clamped at pleasure at any point within the range.

The cross-wire screws are concealed under a little thin brass ferule, placed on the enlarged part of the tube, which can be slipped off, whenever the adjustment of the wires is required.

The cross-wires are adjusted like those of the ordinary level, by revolving the tube in the wyes, and bringing each wire to reverse upon the same object.

When thus adjusted, the telescope is turned in the wyes, until the pinion is in the position shown in the cut, and the wires made respectively horizontal and vertical, when it is clamped by the pins, and the telescope moved up or down at the forward end, and finally clamped upon the sights, so that the vertical wire and the compass sights bisect the same object observed.

The telescope can then be used in place of the sights, and long lines over surfaces nearly horizontal be run with nearly the same accuracy and ease as with the Transit.

When packed in the instrument case, the tube of the telescope takes apart by unscrewing in the middle, there being also little caps supplied to screw on the open ends, and keep out the dust and moisture from the interior of the tube.

MICROMETER.—In this telescope there are three horizontal cross wires, the centre one being fixed as usual, while the others, one on either side, can be adjusted at any distance from each other, and thus made to cover a certain interval upon a rod placed at a specified distance from the telescope.

When these wires are thus adjusted to include a certain

interval, as a foot for instance, upon a rod placed at a distance of 100 feet, it is found that they will cover half a foot at one half that distance, or two feet at a distance of 200 feet, and so on in very nearly the same proportion.

By this means the distance of the rod from the instrument can be measured or set off, without the use of a chain, and with astonishing accuracy and ease.

Indeed, we have been frequently assured that with a powerful telescope, such as we have often placed on our transit instruments, or such as we are now describing, distances can be measured with even greater accuracy than by a chain, especially when the surface of the ground is broken or intersected by deep ravines.

The two small screws by which the movable wires are adjusted, have their heads upon the outside of the washers of the cross wire screws, and can thus be moved by the surveyor with a simple screw driver, until the interval between the wires is made precisely as desired, when the little movable ferule is slipped over all, and the wires protected from any derangement.

When measurements are to be recorded in chains and links, the wires should be made to cover a foot at a distance of 66 feet; if recorded in feet, they should cover the same interval at a distance of 100 feet.

The rod used with the micrometer should be graduated to feet and decimals of a foot, and provided with two targets, the upper one being fixed at some definite point, while the lower one can be moved as the surveyor requires, the distance between the two targets being accurately read off by the vernier of the movable one.

In using the micrometer, the upper wire is brought by the leveling screws of the tripod precisely upon the upper or stationary target, while the lower target is moved up or down until the lower wire exactly bisects its centre line, when the rod is read, and the distance recorded.

THE SOLAR COMPASS.

Fig. 16.

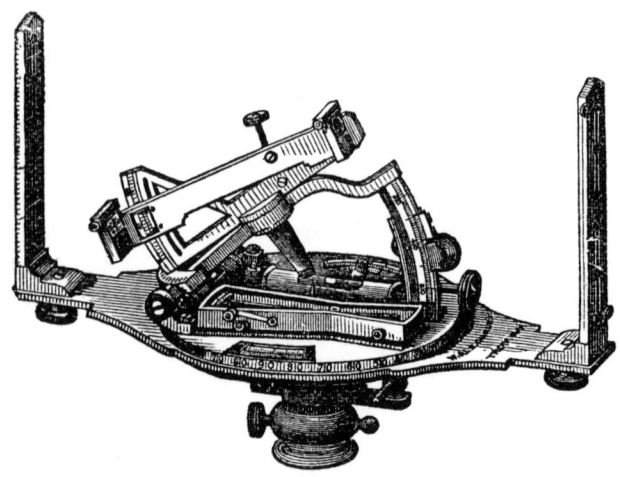

The form of this instrument, which is most commonly used by the government surveyor, is shown in fig. 16, and still better in the engraving at the opening of this article.

As there seen, the main plate which carries the sights, &c., is but a trifle larger than the circular one on which the solar work is placed.

The solar apparatus, which is of the usual form and size, is permanently attached to the sockets of the instrument by a screw, the head of which may be seen in the inside of the socket when the instrument is removed from the ball.

THE SOCKETS are very strong, and the whole instrument is exceedingly compact, light, and substantial.

THE TANGENT MOVEMENT between the plates is partly shown in the figure, the clamp screw, however, is concealed.

THE CLAMP SCREW, by which the instrument is fastened to the spindle, is shown on the side of the socket.

THE SPRING CATCH, of which the knob is shown opposite the head of the clamp screw, prevents the instrument from falling from the spindle when carried on the shoulder, without being previously clamped.

The lines of refraction are shown on one of the sights, but as we have previously remarked, are very seldom used in practice.

THE TANGENT SCALE for angles of elevation and depression is also seen upon the sights, and is sometimes of service in reducing an inclined to a horizontal surface in chaining.

THE GRADUATIONS of this instrument are made upon silver plate, and figured as usual, the arcs and circles being read to single minutes by their respective verniers.

THE SOLAR APPARATUS has the same adjustments and appliances as have been previously described.

THE SOLAR COMPASS should always be furnished with and fitted to a leveling head; but is often used on the compound ball and adjusting socket as shown in Fig. 14, and both of these are now supplied with each instrument.

THE SOLAR TELESCOPE COMPASS.

We have for some years manufactured an instrument named as above. And they have generally performed well, and given satisfaction. But the position of the telescope to one side of the centre, and the excess of weight on that side have always been serious objections to its use in the field, and we have for some time past declined any orders for this instrument, recommending in its place the ordinary form. We have therefore omitted in the present edition the engraving and description of this instrument.

Superiority of our Solar Compasses.

The Solar Compass as hitherto made, though planned with great ingenuity in its general arrangement, was still ex-

tremely rude in its mechanical details and adjustments.
Some of these defects which are apparent on inspection of
any instrument, as hitherto made by other manufacturers,
and which must have frequently occurred to the surveyor,
we will now enumerate.

The motion of the plates over each other was accompanied with so much friction, that in turning the verniers
around the limb, the whole instrument would often be
moved about its spindle.

Again, the verniers must be set, and the sights directed to
an object by the hand alone, a matter of no little difficulty
when single minutes of a degree were to be set off, and accurate observations were required.

The latitude and declination arcs must also be moved by
hand, and the verniers set to single minutes in the same
manner.

The points in which we claim the superiority of our Solar
Compass over any hitherto manufactured, and by means of
which the defects just enumerated are entirely removed,
are partially shown in the various cuts already given, and
will now be stated in detail.

1. A motion of the horizontal plates almost entirely free from friction,
combined with perfect solidity.

2. A fine clamp and tangent movement to the divided limb, as shown
in the figures under the plate.

3. A tangent movement with clamp, for the declination arc, as shown at k.

4. A tangent movement with clamp to the latitude arc, as shown at f.

5. A tangent motion for the whole instrument about its sockets, as
shown in our Adjusting Socket.

6. Great facility of adjustment, and, in consequence, an important saving of time.

7. An important reduction in price, while still furnishing an article
greatly improved.

Weight of the Solar Compass.

Solar Compass, including leveling head....$14\frac{1}{2}$ lbs.

Advantages of the Solar Compass in Surveying.

It will readily occur to all who have read the preceding description of the Solar Compass, that while it is indispensable in the surveys of public lands, it also possesses important advantages over the magnetic compass, when used in the ordinary surveys of farms, &c.

For not only can lines be run and angles be measured without regard to the diurnal variation, or the effect of local attraction, but the bearings being taken from the true meridian, will remain unchanged for all time.

The constant uncertainty caused by the variation of the needle, and the litigation to which it so often gives rise, may thus be entirely prevented by the use of the Solar Compass in this kind of work.

It is also said by those familiar with the use of this instrument, that, in favorable weather, surveys can be more rapidly made with it than with the ordinary needle instrument; there being no time consumed in waiting for the needle to settle, or in avoiding the errors of local attraction.

When the sun is obscured, the lines may be run by the needle alone, it being always kept with the sun, or at 0 on its arc, and thus indicating the direction of the true meridian.

The sun, however, must ever be regarded as the most reliable guide, and should, if possible, be taken at every station.

It is with the design of making the principles and use of the Solar Compass intelligible to the ordinary surveyor, that we have given a more extended account of this instrument than of the others previously mentioned, believing that when its merits become better understood, it will come into more general use.

For much valuable information as to the details of this instrument, as well as the practice of government surveyors in the field, we are indebted to our friend JAMES L. INGALLSBE, late U. S. Deputy Surveyor in Iowa, Kansas and Nebraska.

MEAN REFRACTIONS.

TEMPERATURE 50°, PRESSURE 29.6 INCHES.

From Chambers's Astronomy.

ALTITUDE.	REFRACTION.	ALTITUDE.	REFRACTION.	ALTITUDE.	REFRACTION.	ALTITUDE.	REFRACTION.	ALTITUDE.	REFRACTION.	ALTITUDE.	REFRACTION.
0° 0'	33' 0''	1° 30	21'15''	4° 30'	10'48''	10°30'	5' 0''	25°0'	2' 2''	55°0'	40''
5	32 10	40	20 18	45	10 20	11 0	4 47	27 0	1 51	57 0	37
10	31 22	50	19 25	5 0	9 54	12 0	4 23	29 0	1 42	59 0	34
15	30 36	2 0	18 35	15	9 30	13 0	4 3	31 0	1 35	61 0	32
20	29 50	10	17 48	30	9 8	14 0	3 45	33 0	1 28	63 0	29
25	29 6	20	17 4	45	8 47	15 0	3 30	35 0	1 21	65 0	26
30	28 23	30	16 24	6 0	8 20	16 0	3 17	37 0	1 16	68 0	23
35	27 41	40	15 45	30	7 51	17 0	3 4	39 0	1 10	71 0	19
40	27 0	50	15 9	7 0	7 20	18 0	2 54	41 0	1 5	74 0	16
45	26 20	3 0	14 36	30	6 53	19 0	2 44	43 0	1 1	77 0	13
50	25 42	15	13 39	8 0	6 29	20 0	2 35	45 0	57	80 0	10
55	25 5	30	13 6	30	6 8	21 0	2 27	47 0	53	83 0	7
1 0	24 29	45	12 27	9 0	5 48	22 0	2 20	49 0	49	86 0	4
10	23 20	4 0	11 51	30	5 31	23 0	2 14	51 0	46	89 0	1
20	22 15	15	11 18	10 0	5 15	24 0	2 8	53 0	43	90 0	0

APPROXIMATE EQUATION OF TIME.

From Chambers's Astronomy.

DATE.	MINUTES.	DATE.	MINUTES.	DATE.	MINUTES.	DATE.	MINUTES.
Jan. 1	4	April 1	4	Aug. 9	5	Oct.27	16
3	5	4	3	15	4	Nov.15	15
5	6	7	2	20	3	20	14
7	7	11	1	24	2	24	13
9	8	15	0	28	1	27	12
12	9			31	0	30	11
15	10	19	1			Dec. 2	10
18	11	24	2	Sept. 3	1	5	9
21	12	30	3	6	2	7	8
25	13	May 13	4	9	3	9	7
31	14	29	3	12	4	11	6
Feb. 10	15	June 5	2	15	5	13	5
21	14	10	1	18	6	16	4
27	13	15	0	21	7	18	3
Mch. 4	12			24	8	20	2
8	11	20	1	27	9	22	1
12	10	25	2	30	10	24	0
15	9	29	3	Oct. 3	11		
19	8	July 5	4	6	12	26	1
22	7	11	5	10	13	28	2
25	6	28	6	14	14	30	3
28	5			19	15		

(Column 1: Clock faster than the Sun.) (Column 2: Clock faster. / Clock slower. / Clock faster.) (Column 3: Clock faster. / Clock slower.) (Column 4: Clock slower than the Sun. / Clock faster.)

Made by
W. & L. E. GURLEY,
TROY, N.Y.

SURVEYOR'S TRANSIT
5 inch
Single Vernier. With Level
on Telescope. Vertical Arc
Clamp & tangent to Axis.
and
Patent Solar Attachment

Price as shown above $241.00.

PATENT SOLAR ATTACHMENT,

FOR

TRANSITS.

———————•••———————

Of all the numerous attempts to apply the solar apparatus of Burt to the ordinary transit, there has been nothing devised which, in our opinion, can compare with that shown in the engraving on the preceding page, the patent of which is now owned and controlled by us; and we are ready to supply them to any transit of our own or other make.

In the description of this new attachment, it is supposed that the reader is already familiar with the principles of the solar compass, as well as with the construction and adjustment of the ordinary transit, so fully set forth in our Manual of Instruments, now in the hands of thousands of surveyors in every section of the country.

The solar·attachment we are now considering is essentially the solar apparatus of Burt placed upon the cross bar of the ordinary transit, the polar axis only being directed above instead of below, as in the solar compass.

A little circular disc of an inch and a half diameter, and having a short round pivot projecting above its upper surface, is first securely screwed to the telescope axis.

Upon this pivot rests the enlarged base of the polar axis, which is also firmly connected with the disc by four capstan head screws passing from the under side of the disc into the base already named.

These screws serve to adjust the polar axis, as will be explained hereafter.

The HOUR CIRCLE surrounding the base of the polar axis is easily movable about it, and can be fastened at any point desired by two flat head screws above. It is divided to ten minutes of time; is figured from I. to XII., and is read by a small index fixed to the declination circle, and moving with it.

A hollow cone, or socket, fitting closely to the polar axis and made to move snugly upon it, or clamped at any point desired by a milled head screw on top, furnishes by its two expanded arms below, a firm support for the declination arc, which is securely fastened to it by two large screws, as shown.

The DECLINATION ARC is made larger than in the ordinary solar compass, being of six inches radius, is divided to quarter degrees, and reads by its vernier to thirty seconds of arc, the divisions of both vernier and limb being in the same plane.

The declination arm has the usual lenses and silver plates on the two opposite blocks, made precisely like those of the ordinary solar compass, but its vernier is outside the block, and more easily read.

The declination arm has also a clamp and tangent movement, as shown in the cut. The arc of the declination limb is turned on its axis and one or the other solar lens used, as the sun is north or south of the equator; the cut shows its position when it is south.

THE LATITUDE is set off by means of a large vertical limb having a radius of three inches; the arc is divided to twenty minutes is figured from the centre, each way, up to 80°, and is read by its vernier to thirty seconds.

It has also a clamp screw inserted near its centre, by which it can be set fast to the telescope axis in any desired position.

The vernier of the vertical limb is made movable by the tangent screw attached, so that its zero and that of the limb are readily made to coincide when in adjusting the limb to the level of the telescope, the arc is clamped to the axis.

The usual tangent movement to the telescope axis serves, of course, to bring the vertical limb to the proper elevation, as hereafter described.

A level on the under side of the telescope, with good ground vial and scale, is indispensable in the use of the Solar attachment.

The divided arcs, verniers, and hour circle, are all on silver plate, and are thus easily read and preserved from tarnishing.

The Adjustments.

(1.) THE SOLAR LENSES AND LINES are adjusted precisely like those of the ordinary Solar, the arm being removed and reversed by the opposite faces of the blocks upon the adjuster furnished with each instrument, until the image will remain in the centre of the equatorial lines. This adjustment is very rarely needed in our instruments, the lenses being cemented in their cells, and the plates securely fastened.

(2.) THE VERNIER OF THE DECLINATION ARC is adjusted by setting the vernier at zero, and then raising or lowering the telescope by the tangent screw until the sun's image appears exactly between the equatorial lines.

Having the telescope axis clamped firmly, carefully revolve the arm until the image appears on the other plate.

If precisely between the lines, the adjustment is complete;

if not, move the declination arm by its tangent screw, until the image will come precisely between the lines on the two opposite plates; clamp the arm and remove the index error by loosening two flat head screws on the back, which fasten the movable arc to the declination limb; place the zero of the limb and vernier into exact coincidence and the adjustment is finished.

(3.) To ADJUST THE POLAR AXIS.—First level the instrument carefully by the long level of the telescope, using in the operation the tangent movement of the telescope axis in connection with the leveling screws of the parallel plates until the bubble will remain in the centre during a complete revolution of the instrument upon its axis.

Place the equatorial sights on the top of the blocks as closely as is practicable with the distinct view of a distant object; and having previously set the declination arm at zero, sight through the interval between the equatorial sights and the blocks at some definite point or object, the declination arm being placed over either pair of the capstan head screws on the under side of the disc.

Keeping the declination arm upon the object with one hand, with the other turn the instrument half around on its axis, and sight upon the same object as before. If the sight strikes either above or below, move the two capstan head screws immediately under the arm, loosening one and tightening the other as may be needed until half the error is removed.

Sight again and repeat the operation, if needed, until the sight will strike the same object in both positions of the instrument, when the adjustment of the axis in one direction will be complete.

Now turn the instrument at right angles, keeping the sight still upon the same object as before; if it strikes the same point when sighted through, the axis will be truly vertical in the second position of the instrument.

If not, bring the sight upon the same point by the other pair of capstan head screws now under the declination arc, reverse as before and continue the operation until the same object will keep in the sight in all positions, when the polar axis will be made precisely at right angles to the level and to the line of collimation of the transit.

It should here be noted that as this is by far the most delicate and important adjustment of the solar attachment, it should be made with the greatest care, the bubble kept perfectly in the centre and frequently inspected in the course of the operation.

(4.) To ADJUST THE HOUR ARC.—Whenever the instrument is set in the meridian, as will be hereafter described, the index of the hour arc should read apparent time.

If not, loosen the two flat head screws on the top of the hour circle, and with the hand turn the circle around until it does, fasten the screws again, and the adjustment will be complete.

To obtain mean time, of course the correction of the equation for the given day, as given in the Nautical Almanac, must always be applied.

To find the Latitude.

First level the instrument very carefully, using, as before, the level of the telescope until the bubble will remain in the centre during a complete revolution of the instrument, the tangent movement of the telescope being used in connection with the leveling screws of the parallel plates, and the axis upon the telescope firmly clamped.

Next clamp the vertical arc so that its zero and that of its vernier coincide as near as may be, and then bring them into exact line by the tangent screw of the vernier.

Then, having the declination of the sun for 12 o'clock of the given day as affected by the meridional refraction carefully set off upon the declination arc, note also the equation of time and fifteen or twenty minutes before noon, the telescope being directed to the north, and the object-end lowered until, by moving the instrument upon its spindle and the declination arc from side to side, the sun's image is brought nearly into position between the equatorial lines; now bring the declination arc directly in line with the telescope, clamp the axis firmly, and with the tangent screw bring the image precisely between the lines and keep it there with the tangent screw, raising it as long as it runs below the lower equatorial line, or in other words, as long as the sun continues to rise in the heavens.

When the sun reaches the meridian the image will remain stationary for an instant and then begin to *rise* on the plate.

The moment the image ceases to rise is of course apparent noon, when the index of the hour arc should indicate XII, and the latitude be determined by the reading of the vertical arc.

It must be remembered, however, that the angle through which the polar axis has moved in the operation just described is measured from the zenith instead of the horizon as in the ordinary solar, so that the angle read on the vertical limb is the complement of the latitude.

The latitude itself is readily found by subtracting this angle from 90°; thus at Troy the reading of the limb being found as above directed to be 47° 16' the latitude will be 90°—47° 16'=42° 44'.

It will be noticed that with this apparatus the latitude of any place can be most easily ascertained without any index error, as in the usual solar compass.

To use the Solar Attachment.

From the foregoing description it will be readily understood that good results can not be obtained from the solar attachment unless the transit is of good construction—furnished with the appliances of a level on telescope, clamp and tangent movement to axis, and vertical arc with adjustable vernier, and the sockets or centres in such condition that the level of the telescope will remain in the centre when the instrument is revolved upon either socket.

To run lines with the Solar Attachment.

Having set off the latitude of the place and the declination for the given day and hour, as in the solar, the instrument being also carefully leveled by the telescope bubble, set the horizontal limb at zero and clamp the plates together, loosen the lower clamp so that the transit moves easily upon its lower socket, set the instrument approximately north and south, the object end of the telescope pointing to the north, turn the proper solar lens to the sun, and with one hand on the plates and the other on the revolving arm, move them from side to side until the sun's image is brought between the equatorial lines on the silver plates.

The lower clamp of the instrument should now be fastened and any further lateral movement be made by the tangent screw of the tripod. The necessary allowance being made for refraction, the telescope will be in the true meridian, and being unclamped, may be used like the sights of the ordinary solar compass, but with far greater accuracy and satisfaction in establishing meridian lines. Of course when the upper or vernier plate is unclamped from the limb, any angle read by the verniers is an angle from the meridian, and thus parallels of latitude or any other angles from the true meridian may be established as with the solar compass.

The bearing of the needle, when the telescope is on the meridian, will also give the variation of the needle at the point of observation.

If the instrument, as in our surveyor's transits, has a movable compass circle, the variation of the needle can be set off to single minutes, the needle kept at zero, or "with the sun," and thus lines be run by the needle alone when the sun is obscured.

The variation circle is also applied to engineer's transits of our make, when desired at the time of ordering the same, and without extra charge.

Advantages of the Solar Attachment.

From what has been already said the intelligent surveyor will readily understand that the more perfect horizon obtained by the use of the telescope level, the greater length of the arcs allowing finer readings of angles, and the use of a telescope in place of sights, all render the new attachment more accurate than the ordinary solar compass.

It can also be put on the telescope of any good transit at comparatively small cost, and thus enable the surveyor to establish the true meridian, to determine the correct latitudes, and to obtain true time very nearly.

Its adaptation to the purposes of illustration and instruction in practical astronomy in colleges and schools, will occur to every teacher; and we believe that for the government surveyor it will furnish a long-sought and much needed instrument, superior, in many respects, to the solar compass now so commonly used.

In an experiment just made by us, June 27, 1874, an error of one-quarter of a minute in the direction of the true meridian, or in latitude, could be easily detected by observing the sun's image by a magnifier, and we feel confident that any one who uses the new solar will be surprised and delighted with its work. When desired it can be removed from the telescope and packed in a separate box furnished with it, which also is fitted in the instrument case.

A thin sheath is put on over and protects the polar axis, and is kept in its place by the screw and washer of the socket.

The weight of the new solar attachment is but little over a pound, and is so distributed as not to disturb the counterpoise of the instrument, thus obviating the objection which has hitherto prevented the successful application of the telescope to the solar apparatus.

It is evident that all transits to which the solar attachment is to be applied should have a horizontal limb and verniers, and be leveled by leveling screws and parallel plates.

It can, however, be put on the telescope of our vernier transit compass, but in that case the angles taken from the meridian will be measured by the needle only.

Of course it will be understood, in all cases, that where transits of any kind are to be supplied with the new solar attachment, they must be in perfect order, especially in respect to the sockets, before correct work can be done.

P R I C E S :

Solar attachment, as shown in the cut,......................	$60 00
Vertical arc, with adjustable vernier and tangent screw,........	20 00
Level on telescope, with scale,............................	14 00
Clamp and tangent movement to telescope axis.....	7 00
Total.	$101 00

The price above named includes the cost of putting on and adjusting the new solar attachment to the telescope of a plain transit of any description. Where any of the above appliances or extras are already on the instrument, their price will be deducted from the above.

For a 4½-inch vertical circle of our make, which may be on any transit, and in good order, an allowance of $7.00 will be made.

As shown in the engraving, the cost of the combined transit and solar attachment is $241—made up of the following items:

5-inch single vernier surveyors' transit, plain telescope,.......	$140 00
EXTRAS:	
Level on telescope, with scale,.............................	14 00
Clamp and tangent movement to axis of telescope...........	7 00
Vertical arc with vernier adjusted by tangent screw, divided on silver, and reading to 30 seconds,.....................	20 00
Patent solar attachment, ...	60 00
Total,..	$241 00

The cost of the solar attachment, combined with our double vernier surveyors' transit, and with the same extras, will be $266.00.

Combined with the engineers' transit, and having a variation plate, $281.00.

Where the variation plate is desired in the application of the new solar attachment to any engineer's transit sent to us for the purpose, a charge of $10.00 will be made for the same.

Weight of solar attachment about 18 oz.

ENGINEER'S TRANSIT.

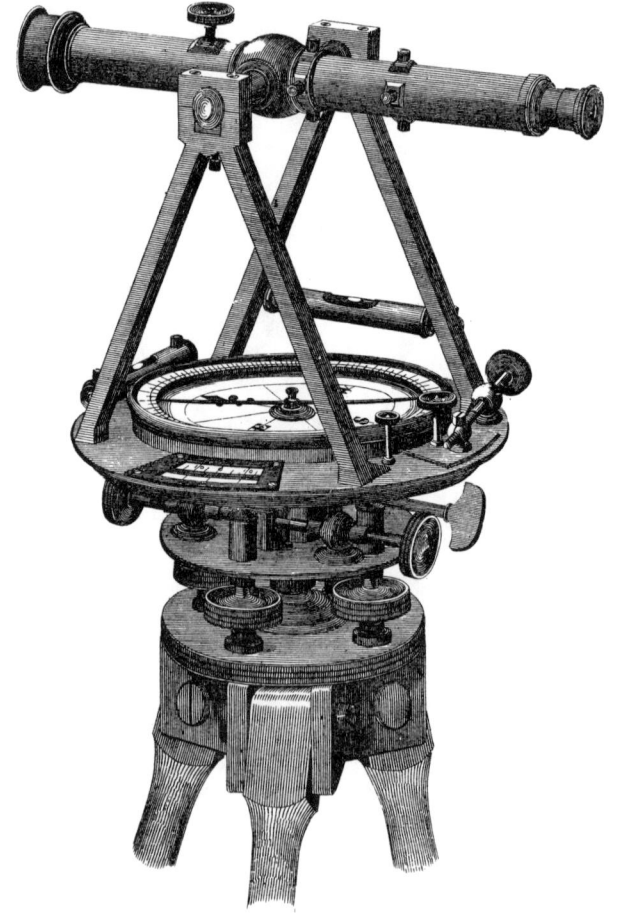

Price as shown above $ 180,00.

Made by

W. & L. E. GURLEY,

TROY, N. Y.

BENJ. D. BENSON, N.Y.

Engineers' Instruments.

THE ENGINEER'S TRANSIT.

Having now described the various instruments employed in surveying, we shall consider those whose use belongs more especially to the practice of the civil engineer, and of these the first in importance is that termed the Engineer's Transit.

The engraving will convey a good idea of our latest improved Engineer's Transit, and to this the reader will please refer in the following detailed description of its different parts.

THE TELESCOPE is from eleven to twelve inches long, and is of the finest quality.

Like those of our other instruments, it is capable of reversion always at the eye end, and we now most commonly make both ends to reverse.

The rack and pinion movement of the object-glass is usually placed, as shown, on the side of the telescope tube, though sometimes on the top, as the engineer may prefer.

PINION TO THE EYE-GLASS.—We have often adapted to the eye-piece of this and our other Transits a rack and pinion movement, which is placed on the side of the tube, and is very excellent in bringing the cross-wires precisely into focus.

A spiral adjustment of the eye-piece is also used by us in the telescopes of all our transits, by which, when the milled head of the eye-piece is twisted in either direction, as may be needed, the eye-piece is brought into focus with ease and accuracy.

THE SHADE.—A short piece of thin tube called the shade, is always made to accompany this and the previous instruments, and is used to protect the object-glass from the glare of the sun, or from moisture ; it must be removed whenever the telescope is reversed, unless the telescope is made to reverse at the eye-end, as is generally desired.

The interior construction of the telescope is similar to those already described.

THE STANDARDS are made of well-hammered brass, firm and strong.

On one of them will be seen the little movable box with the capstan head screw underneath, by which the cross-wires are adjusted to trace a vertical line, as described on page 42 in our account of the Vernier Transit.

THE LIMB or divided circle is seven inches in diameter, graduated to half degrees, and read by two opposite verniers to single minutes.

THE VERNIERS are double, reading both ways from the centre, and are placed on the sides of the plate at right angles to the telescope.

THE NEEDLE is five inches long, and is raised by a milled screw head shown in the cut, placed above the plate.

THE CLAMP AND TANGENT SCREWS are also above, so as to be very accessible, and out of the reach of ordinary accidents. The clamping of the limb is effected in the interior, the aperture being covered with a washer to exclude the dust and moisture.

THE LEVELS, as shown in the cut, are above ; they are both adjustable with the ordinary steel pin.

The glass vials used in the levels of this and the Surveyor's Transit, are ground on their upper interior surface, so that the bubble moves very evenly and with great sensitiveness.

THE TRIPOD HEAD of this instrument is made considerably heavier than that of the Surveyor's Transit.

The upper plate is about five inches diameter, made thick and of well hammered brass ; into this are screwed the long nuts or sockets for the leveling screws, and on the upper surface is seen the clamp, with the two butting tangent screws.

With these the movement is made very slowly, and much more firmly than is possible with a single tangent screw.

The leveling screws are of bell metal, and have a broad three milled head ; they rest on the lower plate, in the little cups spoken of in our account of the previous instrument.

In the engraving it will also be seen that the screws are entirely covered above the plate, by little brass caps which protect the threads from dust and corrosion.

The lower plate is a little smaller than the upper, milled on the edge, and made to connect by a screw, with the tripod legs.

This tripod head is attached to the sockets of the limb and vernier plate, and is removed with them, when the instrument is packed in the box for transportation.

The loop for the plumb-bob is connected by a screw to the spindle of the vernier plate, so that it is always suspended from the exact centre of the instrument.

The Attachments of the Transit.

The engraving of the Surveyor's Transit shows the vertical circle of four and a half inches diameter, which is read by a double vernier to minutes, and also the clamp and tangent movement to the axis of the telescope.

These, with the addition of a level on the telescope, are often used with this instrument, though the majority of engineers prefer an instrument with "plain telescope," like that shown in the engraving.

Micrometer.

It is sometimes very convenient in the use of both the Transit and Leveling Instrument, to employ some simple

method of ascertaining the distances of objects without re-
sorting to actual measurements.

This is well effected by what is termed a "Micrometer," by
the French called "Stadia," the construction and use of
which we have already given on pages 96 — 97, in our
account of the Solar Compass.

The two small screws which adjust the moveable wires
are placed on opposite sides of the telescope, and to one
side of the ordinary cross hair screws, and move the wires
to cover a certain interval upon a rod placed at a specified
distance from the telescope.

The micrometer wire is furnished, whenever desired, with
any of our transits, and without additional charge.

To adjust the Transit.

The adjustments of this instrument and its attachments
have been described in our account of those previously
considered.

To use the Engineer's Transit.

But little need be added to what has been already given
in the previous pages.

THE NEEDLE is of service principally as a rough check
upon the readings of the verniers in the measurement of
horizontal angles, any glaring mistake being detected, by
noticing the angles indicated by both, in the different positions of the telescope.

It may also be used as in the compass, to give the direction in which the lines are run, but its employment is only
subsidiary to the general purposes of the Transit.

Sizes and Weights of Engineer's Transits.

We make three different sizes of this instrument, viz:

4 inch, including leveling head, exclusive of tripod legs, weighs 12½ lbs.
4½   "   "   "   "   14 "
5   "   "   "   "   15 "

Weight of the Attachments.

As it may sometimes be desirable to know the weights of the different extras or attachments, often used in this and the other Transits previously described, we here add them in detail.

Ground level tube, with vial complete.. 7½ oz.
Vertical circle, with vernier .... 6 oz.
Clamp and tangent to axis........... 4 oz.

Besides the simple form of the Engineer's Transit, we also make important modifications, which may be desired by the engineer ; a few of these we shall now enumerate.

The Watch Telescope.

A telescope is sometimes attached to a socket, moving in a hollow cylinder which surrounds the lengthened socket of the limb, and is thus capable of moving around under the plates, and of a short vertical motion.

The cylinder which supports it, may be clamped firmly to he limb, and the wires of the telescope thus fixed upon any object, by the tangent movement of the tripod head.

The object of the watch telescope, is to guard against, and detect any inaccuracy arising from the disturbance of the limb, during the progress of an observation, or the measurement of angles.

Thus, if the wires of both telescopes are fixed upon the same object, and the watch telescope kept still upon it, while the vernier plate is unclamped, and the upper telescope shifted to the second point, a reference to the watch telescope will immediately betray any disturbance in the position of the limb.

But, in spite of its excellencies in cases where great nicety is required, the additional weight and complication of the watch telescope, have caused it to be regarded by most American engineers as an incumbrance, rather than an advantage to the Transit.

The Theodolite Axis.

In place of the ordinary axis of the telescope represented in our engraving, we sometimes make one resembling the Y axis of the English Theodolite.

This modification is desirable, in cases where this instrument is intended to subserve the purposes of both level and transit.

In such an arrangement, the telescope is confined in the axis with clips, by loosening which, it may be revolved in the wyes, or taken out and reversed end for end, precisely like that of the leveling instrument.

The standards also allow its transit, or complete revolution in a vertical direction.

In such an instrument, the adjustment of the wires, and level of the telescope, is effected in the same manner as those of the leveling instrument, the tangent movement of the axis serving, instead of the leveling screws, to bring the bubble and wires into position.

With this modification of the Transit, we have also frequently added, that of a small level bar, wyes, &c., into which the telescope may be transferred, making thus a miniature leveling instrument.

This may be placed upon the socket and tripod head of the transit, and thus made capable of taking levels with a good degree of accuracy.

When desirable, a vertical wheel may be placed on the axis of the telescope of this instrument, and thus all the properties of the English Theodolite united with those of the American Transit

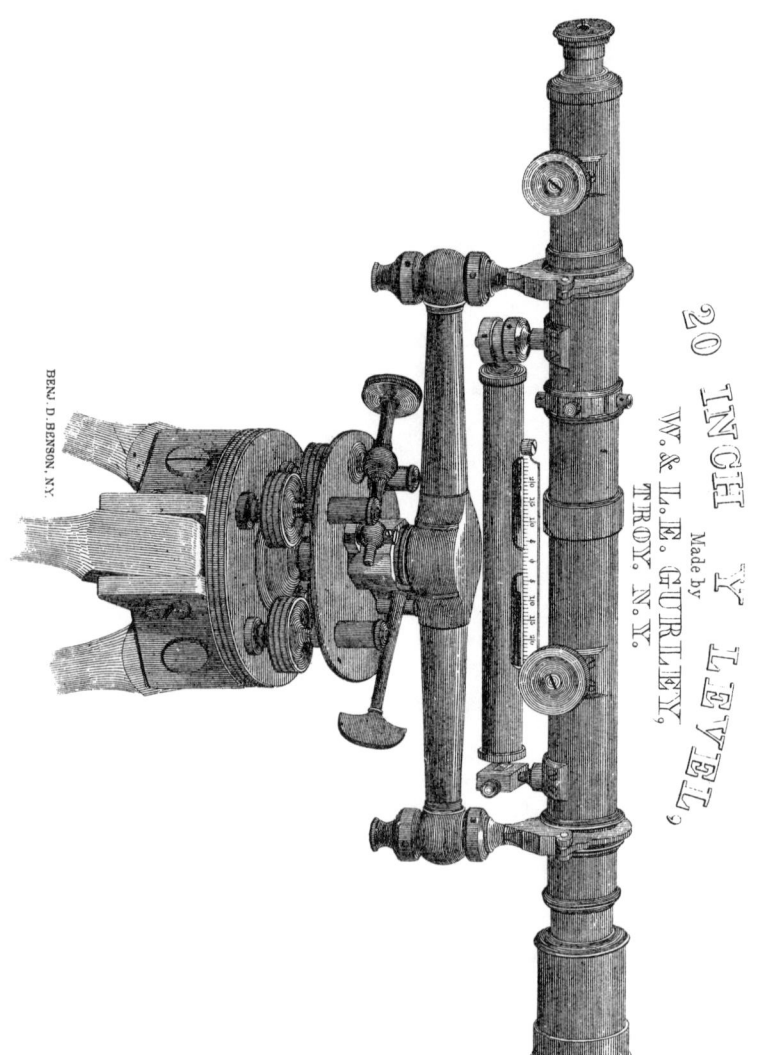

20 INCH Y LEVEL,

Made by

W. & L. E. GURLEY,

TROY, N. Y.

BENJ. D. BENSON, N.Y.

Price, as shown above, $135.00.

THE LEVELING INSTRUMENT.

Of the different varieties of the leveling instrument, that termed the Y Level, has been almost universally preferred by American engineers, on account of the facility of its adjustment and superior accuracy.

Of these levels we manufacture four different sizes, having telescopes of sixteen, eighteen, twenty, and twenty-two inches long, respectively.

The engraving on the opposite page represents our twenty inch Y Level.

We shall consider the several parts of the instrument in detail :

THE TELESCOPE has at each end a ring of bell-metal, turned very truly and both of exactly the same diameter ; by these it revolves in the wyes, or can be at pleasure clamped in any position when the clips of the wyes are brought down upon the rings, by pushing in the tapering pins.

The telescope has a rack and pinion movement to both object and eye-glasses, an adjustment for centering the eye-piece, shown at **A** A, in the longitudinal section of the telescope, (page 105,) and another seen at C, C, for ensuring the accurate projection of the object-glass, in a straight line.

Both of these are completely concealed from observation and disturbance by a thing ring which slides over them.

The telescope has also a shade over the object-glass, so made, that whilst it may be readily moved on its slide over the glass, it cannot be dropped off and lost.

The interior construction of the telescope will be readily understood from fig. 21, which represents a longitudinal section, and exhibits the adjustment which ensures the accurate projection of the object-glass slide.

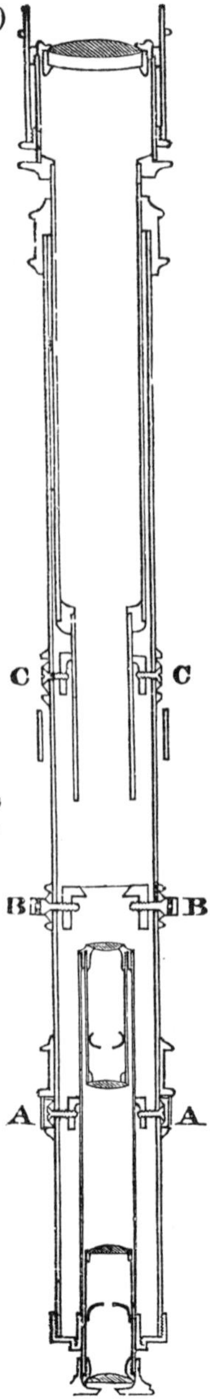

Fig. 21.

As this is peculiar to our instruments, and is always made by the maker so permanently as to need no further attention at the hands of the engineer, we shall here describe the means by which it is effected, somewhat in detail.

The necessity for such an adjustment will appear, when we state, that it is almost impossible to make a telescope tube, so that it shall be perfectly straight on its interior surface.

Such being the case, it is evident that the object-glass slide which is fitted to this surface, and moves in it, must partake of its irregularity, so that the glass and the line of collimation depending upon it, though adjusted in one position of the slide will be thrown out when the slide is moved to a different point.

To prove this, let any level be selected which is constructed in the usual manner, and the line of collimation adjusted upon an object taken as near as the range of the slide will allow; then let another be selected, as distant as may be clearly seen; upon this revolve the wires, and they will almost invariably be found out of adjustment, sometimes to an amount fatal to any confidence in the accuracy of the instrument. The arrangement adopted by us to correct this imperfection, and which so perfectly accomplishes its purpose, is shown in the adjoining cut, fig. 21.

Here are seen the two bearings of the

object-glass slide, one being in the narrow bell-metal ring, which slightly contracts the diameter of the main tube, the other in the small adjustable ring, also of bell-metal, shown at C C, and suspended by four screws in the middle of the telescope.

Advantage is here taken of the fact, that the rays of light are converged by the object-glass, so that none are obstructed by the contraction of the slide, except those which diverge, and which ought always to be intercepted, and absorbed in the blackened surface of the interior of the slide

Now, in such a telescope, the perfection of movement of the slide, depends entirely upon its exterior surfaces, at the points of the two bearings.

These surfaces are easily and accurately turned, concentric, and parallel with each other, and being fitted to the rings, it only remains necessary to adjust the position of the smaller ring, so that its centre will coincide with that of the optical axis of the object-glass.

When this has been once well done, no further correction will be necessary, unless the telescope should be seriously injured.

The manner in which the adjustment of the object-glass slide is effected, will be considered when we come to speak of the other adjustments.

RACK AND PINION.—As seen in the engraving, our Level telescopes are usually furnished with the ordinary rack and pinion movement to both object and eye tubes.

The advantages of an eye-piece pinion, are, that the eye-piece can be shifted without danger of disturbing the tele-scope, and that the wires are more certainly brought into distinct view, so as to avoid effectually any error of observation, arising from what is termed the instrumental parallax.

The position of the pinion on the tube is varied in dif ferent instruments according to the choice of the engineer

We usually place our object slide pinion on the side—both of Transit telescopes, and of those of the Level. The pinion of the eye tube is always placed on the side of the tele scope.

THE LEVEL or ground bubble tube is attached to the under side of the telescope, and furnished at the different ends with the usual movements, in both horizontal and vertical directions.

The aperture of the tube, through which the glass vial appears, is about five and one-fourth inches long, being crossed at the centre by a small rib or bridge, which greatly strengthens the tube.

The level scale which extends over the whole length, is graduated into spaces a little coarser than tenths of an inch, and figured at every fifth division, counting from zero at the centre of the bridge ; the scale is set close to the glass.

The bubble vial is made of thick glass tube, selected so as to have an even bore from end to end, and finely ground on its upper interior surface, that the run of the air bubble may be uniform throughout its whole range.

The sensitiveness of a ground level, is determined best by an instrument called a level tester, having at one end two Y's to hold the tube, and at the other a micrometer wheel divided into hundredths, and attached to the top of a fine threaded screw which raises the end of the tester very gradually.

The number of divisions passed over on the perimeter of the wheel, in carrying the bubble over a tenth of the scale, is the index of the delicacy of the level. In the tester which we use, a movement of the wheel ten divisions to one of the scale, indicates the degree of delicacy generally preferred for railroad engineering.

For canal work practice, a more sensitive bubble is often

desired, as for instance, one of seven or eight divisions of the wheel, to one of the scale.

THE WYES of our levels are made large and strong, of the best bell-metal, and each have two nuts, both being adjustable with the ordinary steel pin.

The clips are brought down on the rings of the telescope tube by the Y pins, which are made tapering, so as to clamp the rings very firmly.

THE LEVEL BAR is made round, of well hammered brass, and shaped, so as to possess the greatest strength in the parts most subject to sudden strains.

Connected with the level bar is the head of the tripod socket.

THE TRIPOD SOCKET is compound; the interior spindle, upon which the whole instrument is supported, is made of steel, and nicely ground, so as to turn evenly and firmly in a hollow cylinder of bell-metal ; this again, has its exterior surface fitted and ground to the main socket of the tripod head.

The bronze cylinder is held upon the spindle by a washer and screw, the head of this having a hole in its centre, through which the string of the plumb bob is passed.

The upper part of the instrument, with the socket, may thus be detached from the tripod head ; and this, also, as in the case of all our instruments, can be unscrewed from the legs, so that both may be conveniently packed in the box.

A little under the upper parallel plate of the tripod head, and in the main socket, is a screw which can be moved into a corresponding crease, turned on the outside of the hollow cylinder, and thus made to hold the instrument in the tripod, when it is carried upon the shoulders.

It will be seen from the engraving, that the arrangement just described allows long sockets, and yet brings the whole instrument down as closely as possible to the tripod head, both objects of great importance in the construction of any instrument.

THE LEVELING HEAD has the same plates and leveling screws as that described in the account of the Engineer's Transit; the tangent screw, however, is commonly single.

For our sixteen inch level we make a similar tripod head, resembling that used with the lighter Engineer's Transit.

The Adjustments.

Having now completed the description of the different parts of the Leveling Instrument, we are ready to proceed with their adjustments, and shall begin with that of the object-slide, which, although always made by the maker, so permanently as to need no further attention at the hands of the engineer, unless in cases of derangement by accident, is yet peculiar to our instruments, and therefore not familiar to many engineers.

To ADJUST THE OBJECT SLIDE.—The maker selects an object as distant as may be distinctly observed, and upon it adjusts the line of collimation, in the manner hereafter described, making the centre of the wires to revolve without passing either above or below the point or line assumed.

In this position, the slide will be drawn in nearly as far as the telescope tube will allow.

He then, with the pinion head, moves out the slide until an object, distant about ten or fifteen feet, is brought clearly into view ; again revolving the telescope in the Y's, he observes whether the wires will reverse upon this second object.

Should this happen to be the case, he will assume, that as the line of collimation is in adjustment for these two distances, it will be so for all intermediate ones, since the bearings of the slide are supposed to be true, and their planes parallel with each other.

If, however, as is most probable, either or both wires fail

to reverse upon the second point, he must then, by estimation, remove half the error by the screws C C, (fig. 21,) at right angles to the hair sought to be corrected, remembering, at the same time, that on account of the inversion of the eye-piece, he must move the slide in the direction which apparently increases the error. When both wires have thus been treated in succession, the line of collimation is adjusted on the near object, and the telescope again brought upon the most distant point; here the tube is again revolved, the reversion of the wires upon the object once more tested, and the correction, if necessary, made in precisely the same manner.

He proceeds thus, until the wires will reverse upon both objects in succession; the line of collimation will then be in adjustment at these and all intermediate points, and by bringing the screw heads, in the course of the operation, to a firm bearing upon the washers beneath them, the adjustable ring will be fastened so as for many years to need no further adjustment.

When this has been completed, the thin brass ferule is slipped over the outside ring, concealing the screw heads, and avoiding the danger of their disturbance by an inexperienced operator.

In effecting this adjustment, it is always best to bring the wires into the centre of the field of view, by moving the little screws A A (fig. 21) working in the ring which embraces the eye-piece tube.

Should the engineer desire to make this adjustment, it will be necessary to remove the bubble tube, in order that the small screw immediately above its scale may be operated upon with the screw-driver.

The adjustment we have now given is preparatory to those which follow, and are common to all leveling instruments of recent construction, and are all that the engineer will have to do with in using our own instruments. What is still necessary then is—

1. *To adjust the line of collimation,* or in other words, to bring both wires into the optical axis, so that their point of intersection will remain on any given point, during an entire revolution of the telescope.

2. *To bring the level bubble parallel* with the bearings of the Y rings, and with the longitudinal axis of the telescope.

3. *To adjust the wyes,* or to bring the bubble into a position at right angles to the vertical axis of the instrument.

To ADJUST THE LINE OF COLLIMATION, set the tripod firmly, remove the Y pins from the clips, so as to allow the telescope to turn freely, clamp the instrument to the tripod head, and, by the leveling and tangent screws, bring either of the wires upon a clearly marked edge of some object, distant from one to five hundred feet.

Then with the hand carefully turn the telescope half way around, so that the same wire is compared with the object assumed.

Should it be found above or below, bring it half way back by moving the capstan head screws at right angles to it, remembering always the inverting property of the eye-piece; now bring the wire again upon the object, and repeat the first operation until it will reverse correctly.

Proceed in the same manner with the other wire until the adjustment is completed.

Should both wires be much out, it will be well to bring them nearly correct before either is entirely adjusted.

When this is effected, slip off the covering of the eye-piece centering screws, shown in the sectional view (fig. 21) at A A, and move each pair in succession with a small screwdriver, until the wires are brought into the centre of the field of view.

The inversion of the eye-piece does not affect this operation, and the screws are moved direct.

To test the correctness of the centering, revolve the telescope, and observe whether it appears to shift the position
of an object.

Should any movement be perceived, the centering is not
perfectly effected.

It may here be repeated, that in all telescopes the position
and adjustment of the line of collimation depends upon
that of the object-glass ; and, therefore, that the movement
of the eye-piece does not effect the adjustment of the wires
in any respect.

When the centering has been once effected it remains permanent, the cover being slipped over to conceal and protect
it from derangement at the hands of the curious or inexperienced operator.

To ADJUST THE LEVEL BUBBLE.—Clamp the instrument
over either pair of leveling screws, and bring the bubble
into the centre of the tube.

Now turn the telescope in the wyes, so as to bring the
level tube on either side of the centre of the bar. Should
the bubble run to the end it would show that the vertical
plane, passing through the centre of the bubble, was not
parallel to that drawn through the axis of the telescope
rings.

To rectify the error, bring it by estimation entirely back,
with the capstan head screws, which are set in either side
of the level holder, placed usually at the object end of the
tube.

Again bring the level tube over the centre of the bar, and
adjust the bubble in the centre, turn the level to either side,
and, if necessary, repeat the correction until the bubble will
keep its position, when the tube is turned half an inch or
more, to either side of the centre of the bar

The necessity for this operation arises from the fact, that when the telescope is reversed end for end in the wyes in the other and principal adjustment of the bubble, we are not certain of placing the level tube in the same vertical plane ; and, therefore, it would be almost impossible to effect the adjustment without a lateral correction.

Having now, in great measure, removed the preparatory difficulties, we proceed to make the level tube parallel with the bearings of the Y rings.

To do this, bring the bubble into the centre with the leveling screws, and then, without jarring the instrument, take the telescope out of the wyes and reverse it end for end. Should the bubble run to either end, lower that end, or what is equivalent, raise the other by turning the small adjusting nuts, on one end of the level, until by estimation half the correction is made ; again bring the bubble into the centre and repeat the whole operation, until the reversion can be made without causing any change in the bubble.

It would be well to test the lateral adjustment, and make such correction as may be necessary in that, before the horizontal adjustment is entirely completed.

To Adjust the Wyes.—Having effected the previous adjustments, it remains now to describe that of the wyes, or, more precisely, that which brings the level into a position at right angles to the vertical axis, so that the bubble will remain in the centre during an entire revolution of the instrument.

To do this, bring the level tube directly over the centre of the bar, and clamp the telescope firmly in the wyes, placing it as before, over two of the leveling screws, unclamp the socket, level the bubble, and turn the instrument half way around, so that the level bar may occupy the same position with respect to the leveling screws beneath.

Should the bubble run to either end, bring it half way back by the Y nuts on either end of the bar; now move the telescope over the other set of leveling screws, bring the bubble again into the centre, and proceed precisely as above described, changing to each pair of screws, successively, until the adjustment is very nearly perfected, when it may be completed over a single pair.

The object of this approximate adjustment, is to bring the upper parallel plate of the tripod head into a position as nearly horizontal as possible, in order that no essential error may arise, in case the level, when reversed, is not brought precisely to its former situation. When the level has been thus completely adjusted, if the instrument is properly made, and the sockets well fitted to each other and the tripod head, the bubble will reverse over each pair of screws in any position.

Should the engineer be unable to make it perform correctly, he should examine the outside socket carefully to see that it sets securely in the main socket, and also notice that the clamp does not bear upon the ring which it encircles.

When these are correct, and the error is still manifested, it will, probably, be in the imperfection of the interior spindle.

After the adjustments of the level have been effected, and the bubble remains in the centre, in any position of the socket, the engineer should carefully turn the telescope in the wyes, and sighting upon the end of the level, which has the horizontal adjustments along each side of the wye, make the tube as nearly vertical as possible.

When this has been secured, he may observe, through the telescope, the vertical edge of a building, noticing if the vertical hair is parallel to it; if not, he should loosen two of the crosswire screws at right angles to each other and with the hand on these, turn the ring inside, until the hair is made vertical; the line of collimation must then be corrected again, and the adjustments of the level will be complete.

To use the Level.

When using the instrument, the legs must be set firmly into the ground, and neither the hands nor person of the operator be allowed to touch them; the bubble should then be brought over each pair of leveling screws successively, and leveled in each position, any correction being made in the adjustments that may appear necessary.

Care should be taken to bring the wires precisely in focus, and the object distinctly in view, so that all errors of parallax may be avoided.

This error is seen when the eye of an observer is moved to either side of the centre of the eye-piece of a telescope, in which the foci of the object and eye-glasses are not brought precisely upon the cross-wires and object; in such a case the wires will appear to move over the surface, and the observation will be liable to inaccuracy.

In all instances the wires and object should be brought into view so perfectly, that the spider lines will appear to be fastened to the surface, and will remain in that position however the eye is moved.

If the socket of the instrument becomes so firmly set in the tripod head as to be difficult of removal in the ordinary way, the engineer should place the palm of his hand under the wye nuts at each end of the bar, and give a sudden upward shock to the bar, taking care also to hold his hands so as to grasp it the moment it is free.

BUILDER'S LEVEL,

15 INCH TELESCOPE

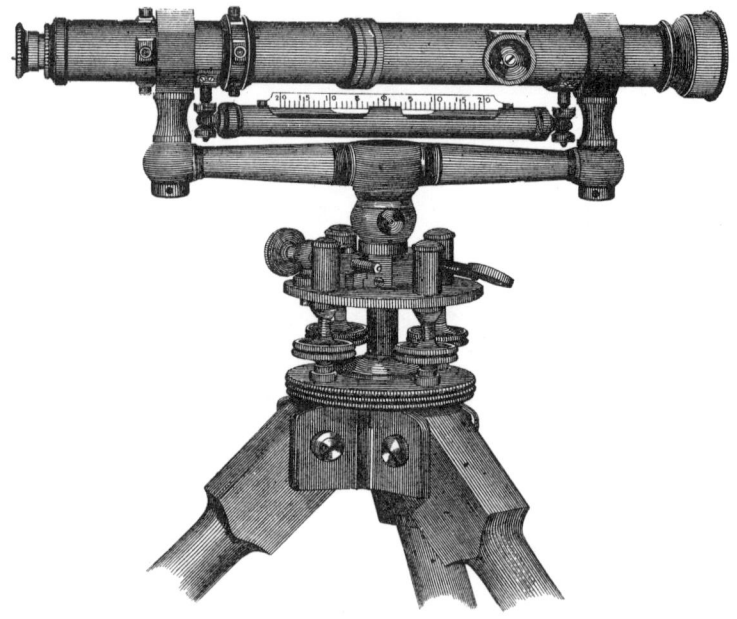

Price as shown above $ 75.00.

Made by

W. & L. E. GURLEY,

TROY, N.Y.

Weight of Leveling Instruments.

The average weights of the different sizes of this instrument, exclusive of the tripod legs, are as follows:

16-inch telescope, with leveling head......11 lbs.
18-inch   "   "   ......$12\frac{1}{2}$ "
20-inch   "   "   ......$13\frac{1}{4}$ "
22-inch   "   "   ......$13\frac{3}{4}$ "

THE BUILDER'S, OR DUMPY LEVEL.

This instrument, one size of which is shown in the en
graving, is of more simple and compact construction than
those already described.

As represented, the telescope is provided with the usual
facilities for adjustment, and rests upon two hollow cylin-
ders or studs, raised from the ends of the bar by the two
similar faces of the octagonal-shaped prisms which sur-
round the telescope tube at either end.

The telescope and attachments are held firmly to the
bar by a long, stout screw at either end; the heads of
these screws are shown under the ends of the bar, and are
bored to admit the usual adjusting-pin.

A strong spiral spring is placed in a recess in the upper
end of the upright stud at each end of the bar, and serves
in connection with the screws to effect the third adjustment of the level.

These springs are, of course, removed whilst the other
adjustments are in progress, and the telescope allowed to
rest directly upon the upper plane surface of the upright
studs.

The level is placed under the telescope like that of the Y
level, and is adjustable at either end by two nuts, as shown.
The instrument should always be used upon the adjusting
tripod with leveling screws, as shown in the engraving.

The adjustments of this instrument are made in the same order and almost precisely in the same manner as those of the Engineers' Level, described on pages 114–118, and need but a brief description here.

To adjust the time of collimation, it is necessary first to remove the two long stout screws which hold the telescope to the bar, and also the springs already named, from the upper ends of the studs, so that the similar faces of the prisms may rest directly upon the upper surfaces of the studs.

(1.) The line of collimation is then adjusted like that of the ordinary Y level, by making the cross wires to reverse upon any given point when the telescope is turned half way around, so as to rest upon the studs by the opposite faces of the prisms.

(2.) The level is adjusted by turning the telescope end for end upon the studs, the bubble being made to come to the centre in both positions.

(3.) The bubble is brought into a position at right angles to the vertical axis (the adjustment of the wyes in ordinary levels), by releasing or compressing the springs at either end by turning the capstan head screws underneath the bar, so that the bubble will come into the centre as the instrument is turned upon its spindle, over both pairs of leveling screws in succession.

This instrument is packed in the box with the leveling head always attached, and need not to be taken from its spindle except for repairs.

We make two sizes of this level—one, of 11-inch telescope, with light tripod, and used by millwrights, masons and builders, and the 15-inch size, with medium-sized tripod—used not only by the persons above named, but found to be a most simple and accurate level in railroad surveys and constructions.

LEVELING RODS.

The three kinds used by American engineers are all sliding rods. The Philadelphia Rod is divided to tenths, and reads to two hundredths of a foot. The Boston and New York are divided into hundredths of a foot and reading by verniers to thousandths.

Philadelphia Rod.

The leveling rod known as the Philadelphia Rod, is formed of two strips of light baywood or mahogany, each three-fourths of an inch by one and three-fourths inches by six and one-half feet long, connected together by two metal sleeves, the upper one of which has a clamping screw, for fixing the rod in its position when the two parts are separated or extended beyond six feet six inches.

Both sides of the back rod and one side of the front rod are planed out one-sixteenth of an inch below the edges. These depressed surfaces are all painted white, and divided into feet, and tenths of a foot. The front rod reads from the foot upward; both sides of the back rod read downward. The feet figures are red, one inch long, and the one-tenth figures black, eight-tenths of an inch long. The target is round, made of brass, with an opening, in its face, two and one-fourth inches long by one inch wide, with a divided scale on one side reading the rod to two hundredths of a foot.

The upper sleeve has a vernier reading to one-hundredths on the back of the rod, for the rod-man to take the reading when the two parts are extended beyond six feet six inches.

No. 90.—Philadelphia Rod.

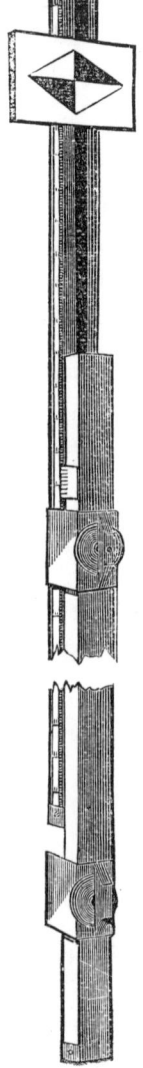

No. 91.—*Boston Rod.*

Boston Rod.

That known as the Boston or Yankee Rod, is formed of two pieces of light bay-wood or mahogany, each about six and a half feet long, connected together by a tongue, and sliding easily by each other, in both directions.

One side is furnished with a clamp screw and vernier at each end, the other carries the divisions, marked on strips of satin wood, inlaid on either side.

The target is a rectangle of wood, fastened near one end of the divided side, and having its horizontal line just three-tenths from the extremity.

The target being fixed, when any height is taken above six feet, the rod is changed end for end, and the divisions read by the other vernier; the height to which the rod can be extended being a little over eleven feet.

This kind of rod is very convenient from its great lightness, but the parts are made too frail to endure the rough usage of this country, and, therefore, American engineers have generally given the preference to another, made heavier and more substantial.

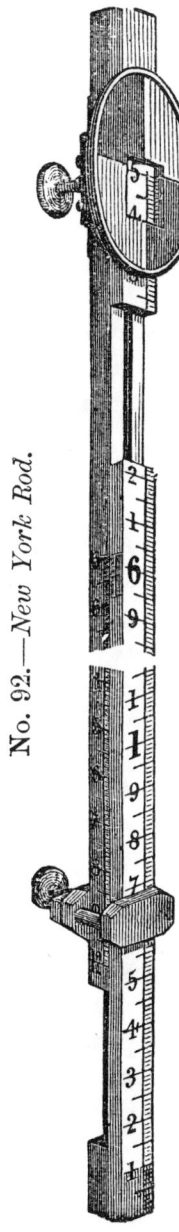

No. 92.—*New York Rod.*

The New York Rod.

This rod, which is shown in the engraving as cut in two, so that the ends may be exhibited, is made of satin wood, in two pieces like the former, but sliding one from the other, the same end being always held on the ground, and the graduations starting from that point.

The graduations are made to tenths and hundredths of a foot, the tenth figures being black, and the feet marked with a large red figure.

The front surface, on which the target moves, reads to six and a half feet; when a greater height is required, the horizontal line of the target is fixed at that point, and the upper half of the rod, carrying the target, is moved out of the lower, the reading being now obtained by a vernier on the graduated side, up to an elevation of twelve feet.

The mountings of this rod are differently made by different manufacturers. We shall give those which we have adopted.

The target is round, made of thick brass, having, to strengthen it still more, a rib raised on the edge, which also protects the paint from being defaced.

The target moves easily on the rod, being kept in any position by the friction of the two flat plates of brass which are pressed against two alternative sides, by small spiral springs, working in little thimbles attached to the band which surrounds the rod.

There is also a clamp screw on the back, by which it may be securely fastened to any part of the rod.

The face of the target is divided into quadrants, by horizontal and vertical diameters, which are also the boundaries of the alternate colors with which it is painted.

The colors usually preferred are white and red: sometimes white and black.

The opening in the face of the target is a little more than a tenth of a foot long, so that in any position a tenth, or a foot figure, can be seen on the surface of the rod.

The right edge of the opening is chamfered, and divided into ten equal spaces, corresponding with nine hundredths on the rod; the divisions start from the horizontal line which separates the colors of the face.

The vernier, like that on the other side of the rod, reads to thousandths of a foot.

The clamp, which is screwed fast to the lower end of the upper sliding piece, has a movable part which can be brought by the clamp screw firmly against the front surface of the lower half of the rod, and thus the two parts immovably fastened to each other without marring the divided face of the rod.

MINERS' COMPASSES.

These instruments shown in the engraving consist essentially of a magnetic needle so suspended as to move readily in a vertical direction, the angle of inclination, or "dip," being measured upon the divided rim of a small compass box.

When in use, the ring or bail is held in the hand—the compass box by its own weight takes a vertical position—and should also be in the plane of the magnetic meridian.

In this position the needle, when unaffected by the attraction of iron, assumes a horizontal line, as shown by the zeros of the circle. When brought over any mass of iron it dips, and thus detects the presence of iron ores with certainty.

MINER'S COMPASSES.

43

"Norwegian"
Glass both sides.
Price
$12.00. & $15.00.

40

Glass both sides.
with stop to needle
Price $12,00.

Made by

W.&.L.E.GURLEY,

TROY, N.Y.

42

Brass back
With stop Price $12,00.

BENJ. D. BENSON, N.Y.

If the compass box of either No. 40 or 42 is held horizontally, it serves as an ordinary pocket compass. The box of the Miner's Compass is made of brass, with cover of same material, or, if so ordered, it is packed in small mahogany case. The form shown in the engraving, No. 40, has its two sides of glass, and is provided with a stop for the needle, worked by the little brass knob there shown.

The compass No. 42 has a brass back and brass cover, and is used as above described, the observer standing with his face to the west, and holding the compass box suspended in the plane of the magnetic meridian.

NORWEGIAN COMPASS.
Glass both sides, brass covers.

Besides the ordinary article just described, we have lately introduced a modification of one used in northern Europe. This has the needle resting upon a single vertical pivot so as to move freely in a horizontal direction, and thus place itself with certainty in the magnetic meridian, while at the same time being attached to the needle cap by two delicate pivots, one on each side, it is free to dip—like that of the ordinary miner's compass, described above.

THE POCKET COMPASS.

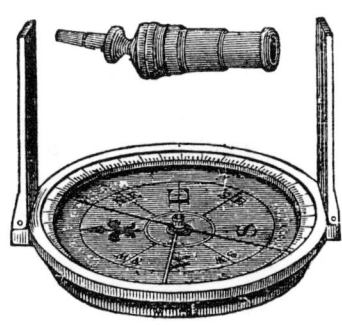

This little instrument, shown with Jacob-staff socket in Fig. 26, though not used in extensive surveys like the larger compasses we have described, is found very convenient in making explorations, or in retracing the lines of government surveys, as in locating land warrants, etc.

Fig. 26.

The sights are made with a slote and a hair, on opposite sides; they also have joints near the base, so as to fold over

The circle is graduated to degrees, and figured from 0 to 90 each way, as in larger instruments.

The needle is suspended upon a jeweled centre, and is raised by the lifter shown in the cut.

The jacob-staff socket is often used with the compass, being screwed to the under side, and detached at pleasure.

The mountings are all that are furnished, the staff itself being easily made out of a common walking-stick.

We make two sizes of the Pocket Compass, differing mainly in the needle, which in one is two and a half, in the other three and a half inches long.

The larger size is also sometimes provided with two small levels, let into the face of the compass ; these are not shown in the cut.

VERNIER POCKET COMPASS.

This instrument, shown in the engraving, has also a three and a half inch needle, and is furnished with a vernier outside, reading to five minutes, by which the sights can be placed at any desired angle with the line of zeros, so as to set off the variation of the needle, as with the Vernier Compass.

The compass is furnished with jacob-staff mountings ; sometimes, if desired, with a very light tripod, as shown in the engraving ; has two levels, and is neatly packed in a mahogany case.

It makes a most excellent and portable little instrument in locations, and is especially useful for the surveyor of government lands.

THE VERNIER POCKET COMPASS.

Made by

W. & L. E. GURLEY,

TROY, N.Y.

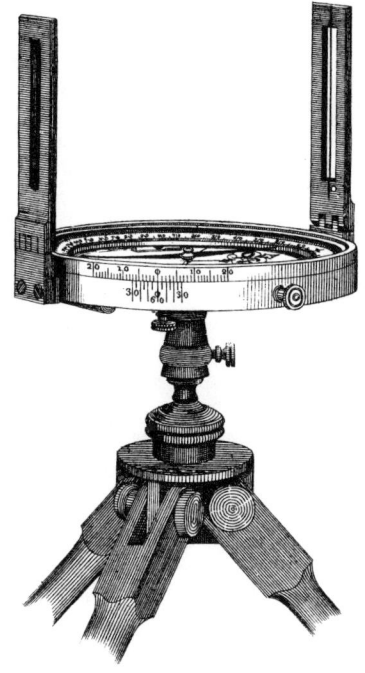

Price as shown above $ 23,00.

General Matters.

TRIPODS.

In the tripods of all our instruments, the upper part of the leg, is flattened, and fitted closely in the surfaces of the brass cheek pieces.

The cheeks are made very broad, and give a firm hold upon the leg, which may be tightened at any time by screwing up the bolts which pass through the top of the legs ; this is especially necessary after the surface of the wood has been much worn.

The legs are round, and taper in each direction from a swell, turned about one-third the way down, from the head to the point.

The point, or shoe, is a tapering brass ferule, having an iron end ; it is cemented, and riveted firmly to the wood.

The legs of all our tripods are about four feet eight inches long, from head to point. We make five sizes of tripods, which we will now separately describe.

1. THE COMPASS TRIPOD, seen in part in the cut of the vernier transit, and having the brass plate to which the cheeks are attached, three and three-fourth inches in diameter, and legs which are about one inch at the top, one and three-eighths at the swell, and seven-eighths at the bottom.

The legs are usually made of cherry, sometimes of mahogany, and the tripod is used with the various kinds of compasses, and with the vernier transit.

2. THE MEDIUM SIZED TRIPOD, shown with the surveyor's transit, and having a plate of same diameter as above, but with the cheeks made considerably broader, by curving at each end ; the legs being also about an eighth of an inch larger throughout.

This tripod has mahogany legs, and is used with the surveyor's transit, the light engineer's transit, and the sixteen inch level.

3. THE HEAVY TRIPOD, shown with the engineer's transit, having a brass plate of four and one-fourth inches diameter, with extended cheek pieces, and with legs one and three-eighths of an inch at the top, one and three-fourths at the swell, and one and an eighth at the point.

The heavy size has also mahogany legs, and is used with the engineer's transit, and larger leveling instruments.

4. SHORT TRIPODS, FOR MINES.—Tripods of one half the usual length, for mining engineering, furnished in place of those usually sent, if so ordered.

5. JOINTED TRIPODS.—When parties so order we put screw joint and point in the centre of each leg of tripods, which enables the engineer to use it for mining purposes ; this improvement adds $5 00 to the cost of instrument or tripod.

Lacquering.

All instruments are covered with a thin varnish, made by dissolving gum shellac in alcohol, and applied when the work is heated.

As long as this varnish remains, the brass surface will be kept from tarnishing, and the engineer, by taking care not to rub his instrument with a dusty cloth, or to expose it to the friction of his clothes, can preserve its original freshness for a long time.

Bronze Finish.

Instead of the ordinary brass finish, most engineers prefer instruments blackened or bronzed. This is done with an acid preparation, after the work has been polished, and gives the instrument a very showy appearance, besides being thought advantageous on account of not reflecting the rays of the sun as much as the ordinary finish.

We finish our instruments either bright or bronze, as may be preferred.

CHAINS.

Surveyors' Chains.

FOUR POLE CHAINS.—The ordinary surveyor's chain is sixty-six feet, or four poles long, composed of one hundred links, each connected to the other by two rings, and furnished with tally marks at the end of every ten links.

In all the chains we manufacture, the rings are oval, are sawed, and well closed, the ends of the wire forming the hook being also filed and bent close to the link, so as to avoid the danger of "kinking."

A link in measurement includes a ring at each end.

The handles are of brass, and each forms part of the end links, to which it is connected by a nut, by which also the length of the chain is adjusted.

The tallies are also of brass, and have one, two, three, or four, notches, as they are ten, twenty, thirty, or forty, links, from either end ; the fiftieth link is rounded, so as to distinguish it from the others.

TWO POLE CHAINS.—In place of the four pole chain just described, many surveyors prefer one of two rods or thirty-three feet long, having but fifty links, and counted by its tallies from one end in a single direction.

SNAP FOR ALTERING CHAINS.—We often make four pole chains so arranged, that by detaching a steel snap in the middle, the two parts can be separated, and then one of the handles being removed in the same manner, and transferred to the forty-ninth link, a two pole chain is readily obtained. This modification is made whenever desired, and without any additional charge.

VARA, METER, AND PENNSYLVANIA CHAINS. *See Price List, pages* 8 *and* 9.

SIZES OF WIRE.—Our surveyors' chains are made of the best refined iron wire, of sizes No. 8 or 10, as may be preferred ; the diameter of No. 10 wire being about one-eighth of an inch, and that of No. 8 wire nearly a sixteenth larger.

Engineers' Chains

Differ from the preceding, in that the links are each 12 inches long ; the wire, also, is usually much stronger.

They are either fifty or one hundred feet long, and are furnished with handles, tallies, &c., and sometimes with a swivel in the middle to avoid being twisted in use.

In place of the round rings commonly made, we have substituted in these, and our other chains, rings of an oval form, and find them almost one-third stronger, though made of the same kind of wire.

SIZES OF WIRE.—The wire used for these chains is commonly of No. 5 or 6 ; the first being nearly one-fourth of an inch in diameter, while No. 6 wire is about one-sixteenth smaller.

The wire is of the first quality, and the whole chain is made in the most accurate and substantial manner.

Steel Chains.

Chains made of steel wire, though more costly than those which we have just described, are yet often preferred on account of their greater lightness and strength.

They are made of any desired size or length, generally of No. 10, rarely of No. 8 wire, and are very stiff and strong.

BRAZED STEEL CHAINS.—A very portable and excellent measure is made, by a light steel chain, each link and ring of which is securely *brazed*, after being united together and tested the wire is also tempered.

The wire generally used by us is of size No. 12, the rings are of oval form, the chain, though exceedingly light, is almost incapable of being either broken or stretched.

Our steel brazed chains have been found exceedingly desirable for all kinds of measurement, and for the use of engineers upon railroads and canals have almost entirely superseded the heavier chains.

Grumman's Patent Chains.

These chains, invented and patented by J. M. Grumman, of Brooklyn, N. Y., are made of very light steel wire, the links being finely tempered, and, as shown in the illustration,

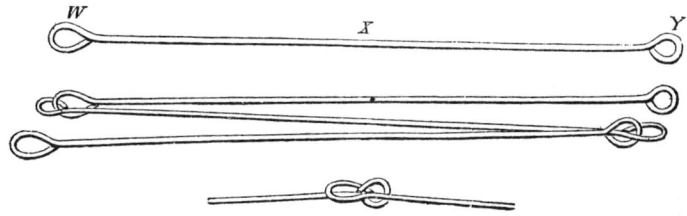

so formed at the ends as to fold together readily, and thus dispense with the use of rings.

This construction gives only one-third as many wearing points as the ordinary chain, and affords the utmost facility for repairs, from five to ten extra links being furnished with each chain, which have only to be sprung into place to replace such as may have been broken; it can also be taken apart at any link, and, by having a spring-catch on either handle, be made of any length desired. These chains are made of three different sizes of wire—the first two, termed drag-chains, being of size No. 12 and 15, and used for measuring on the surface, like the ordinary chain; and the second, called the "suspended chain," for very accurate measurements, made of No. 18 wire, and with spring-balance, thermometer and spirit level attachments, to be held above the surface when in use, the extremities of the chain being marked upon the ground by the points of plummets let fall from the ends of the chain.

The drag-chains are all that are needed in common land surveys ; for a mixed practice of village and country surveying, the spring-balance should be attached to the drag-chains, while for city surveying the suspended chain, with all its attachments, is the proper instrument.

We have purchased the patent for the Grumman chains, with the entire right to make and sell them, and shall hereafter be able to furnish them promptly.

Vara Chains.

The Spanish or Mexican Vara, which is in very general use in Texas, Mexico, Cuba and South America, is $33\frac{1}{3}$ inches long. The chains are made of ten or twenty varas, each vara being usually divided into five links, a link, including a ring at each end, is, therefore, $6\frac{2}{3}$ inches. A chain of ten varas has fifty links ; of twenty varas one hundred links. Each vara is marked by a round brass tally, numbered from one to nine in the ten-vara chain, and from one to ten—each way, in the twenty-vara chain. Sometimes, but rarely, the vara is divided into four links, a ten-vara chain then has forty links, and a twenty vara, eighty links.

Marking Pins.

In chaining, there are needed ten marking pins, or chain stakes, made either of iron, steel, or brass wire, as may be preferred, about fourteen inches long, pointed at one end to enter the ground, and formed into a ring at the other, for convenience in handling.

They are sometimes loaded with a little mass of lead around the lower end, so as to answer as a plumb when dropped to the ground, from the suspended end of the chain.

To use the Chain.

In using the chain its length must be taken from its extreme ends, and the pins placed on the outside of the handles; it must be drawn straight and taut, and carefully examined to detect any kinks or other causes of inaccuracy.

Our chains are all carefully tested at every ten, sometimes at every link, and in their whole length by the U. S. standard, and when new may always be relied upon as correct.

But as all will alter, more or less, after long use in the field, it will be best for the surveyor to carefully lay down on a level surface the exact length of the chain when yet new, marking also its extreme ends by monuments which will not be liable to disturbance.

He will thus have a standard measure of his own to which the chain can be adjusted from time to time, and again be used with perfect confidence.

TAPE MEASURES.

The best are Chesterman's steel tapes, made of a thin ribbon of steel, which is jointed at intervals, and wound up in a leathern case, having a folding handle.

These tapes are of all lengths, from three to one hundred feet divided into inches and links, or more usually, tenths of a foot, and links, the figures and graduations being raised on the surface of the steel.

CHESTERMAN'S METALLIC TAPES.

These are of linen, and have also fine brass wires interwoven through their whole length.

They are thus measurably correct, even when wet.

They are mounted like the steel tapes, of like lengths, and similarly graduated. *For prices see pages* 10 *and* 11.

August 15th, 1874.

Supplement to Manual.

☞ When ordering goods always state what edition of Manual, and number in Catalogue.

₌ The prices in this Catalogue may vary from time to time, on account of fluctuations in Gold Premium and Market Rates.

DRAWING INSTRUMENTS.

To guide the surveyor and engineer in the selection of Drawing Instruments, we here add a detailed description, with illustrations and prices of the separate pieces, and cases of the different kinds in general use.

Those we shall first mention are of Swiss manufacture, and are of the finest quality and finish.

The Brass Instruments are used in Schools and elementary practice.

The fine German Silver Instruments are of the best German make, intended for Engineers, Architects, and Machinists:

Swiss Drawing Instruments.

OF GERMAN SILVER, EXTRA FINE FINISHED.

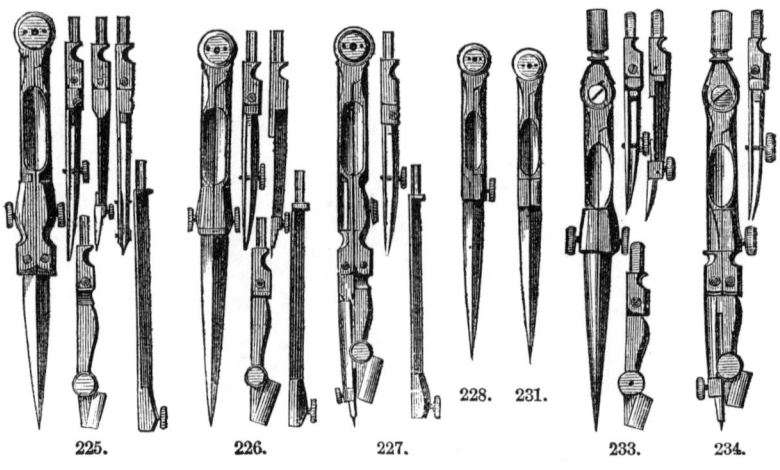

225.   226.   227.   228.   231.   233.   234.

No.   Price.

225.—Drawing Compass, joints in legs, 6½ inches long, with pen, pencil-
        holder, needle point, lengthening bar and dot. pen..................$11 00
226.—Drawing Compass, 6½ inches long, with pen, pencil-holder, lengthening
        bar and needle-point.. 8 50
227.—Drawing Compasses, 6½ inches long, with fixed Needle Point and Loose
        Pen, and Pen Points and Lengthening Bar 6 75
228.—Hair Spring Dividers, 4½ inches long................................... 2 65
229.—   "   "   5 to 6 inches long.............................. 3 00
230.—Plain Dividers, 4½ inch... 1 90
231.—   "   5 inch.. 2 25
232.—   "   6 inch.. 2 70
233.—Drawing Compass, 4 inch, with pen, pencil-holder, and needle-point..... 6 00
234.—Drawing Compass, 4 inch, with fixed needle-point, and pen and pencil-
        point, changeable... 5 25
235.—Proportional Dividers, 6½ inches long, finely graduated for lines........ 8 75
236.—Proportional Dividers, 6½ inches long, finely graduated for lines and poly-
        gons. 10 00
237.—Proportional Dividers, 9 inches long, finely graduated for lines and poly-
        gons... 12 25
238.—Proportional Dividers, 9 inches long, with micrometer adjustment (238),
        finely graduated for lines and polygons............................. 14 65

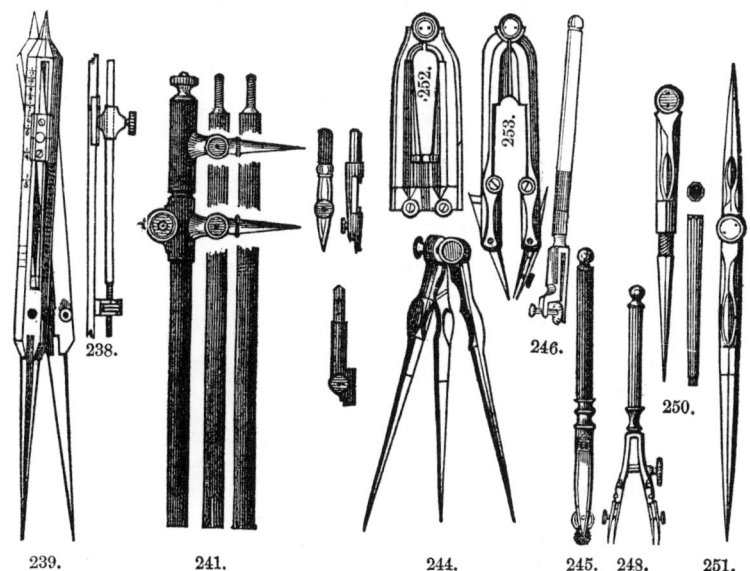

239.   241.   244.   245. 248.   251.

239.—Proportional Dividers, 8 inches long, with rack adjustment, graduated
        for lines... 12 75
240.—Beam Compass, 19-20 inches long, in two German Silver bars........... 11 50

No.	Price.
241.—Beam Compass, 21 inches long, in 3 German Silver bars	13 00
242.—Beam Compass, 36 ,, 4 German Silver Bars	19 00
243.—Beam Compass, 54 " 4 German Silver Bars	22 50
244 —Triangular Compass	5 25
245.—Dotting Pen, with one wheel $2 65, with six wheels	4 00
246.—Dotting Pen, New Style, in case, fine article	3 50
247.—Road or Double Drawing Pen	4 15
248.— " " Joint on each side	3 80
250.—Pocket dividers, with sheath	3 00
251.—Whole and Half Dividers	4 30
252.—Universal Compass, with points to shift	8 50
253.— " " " to TURN	9 00
254.— " " " to change, and handles to bow pen and pencil	10 00

255.   256.   257.   259.   260.   261.

255.—Dividers, 4 inches long, with two fixed Needle Points	3 25
256.—Dividers, 4 inches long, with fixed Needle Point and Pen Point	3 60
257.—Dividers, 4 inches long, with fixed Needle Point and Pencil Point	3 60
258.—Large Steel Spacing, Dividers 5-inch	3 20
259.—Small Steel Spacing Dividers, 3½ inches	1 70
260.—Small Steel Spacing Dividers, 3½ inches long, Ivory Handle and Needle Points	3 00
261.—Small Steel Bow Pen, 3½ inches	2 25
262.—Small Steel Bow Pen, Needle Point	3 00
263.—Small Steel Bow Pencil, 3½ inches	2 25
264.—Small Steel Bow Pencil, with Needle point	3 00
266.—Bow Pen, German Silver	2 65
267.—Bow Pen, with pencil-holder, German Silver	3 60
271.—Eccentric rule	2 65

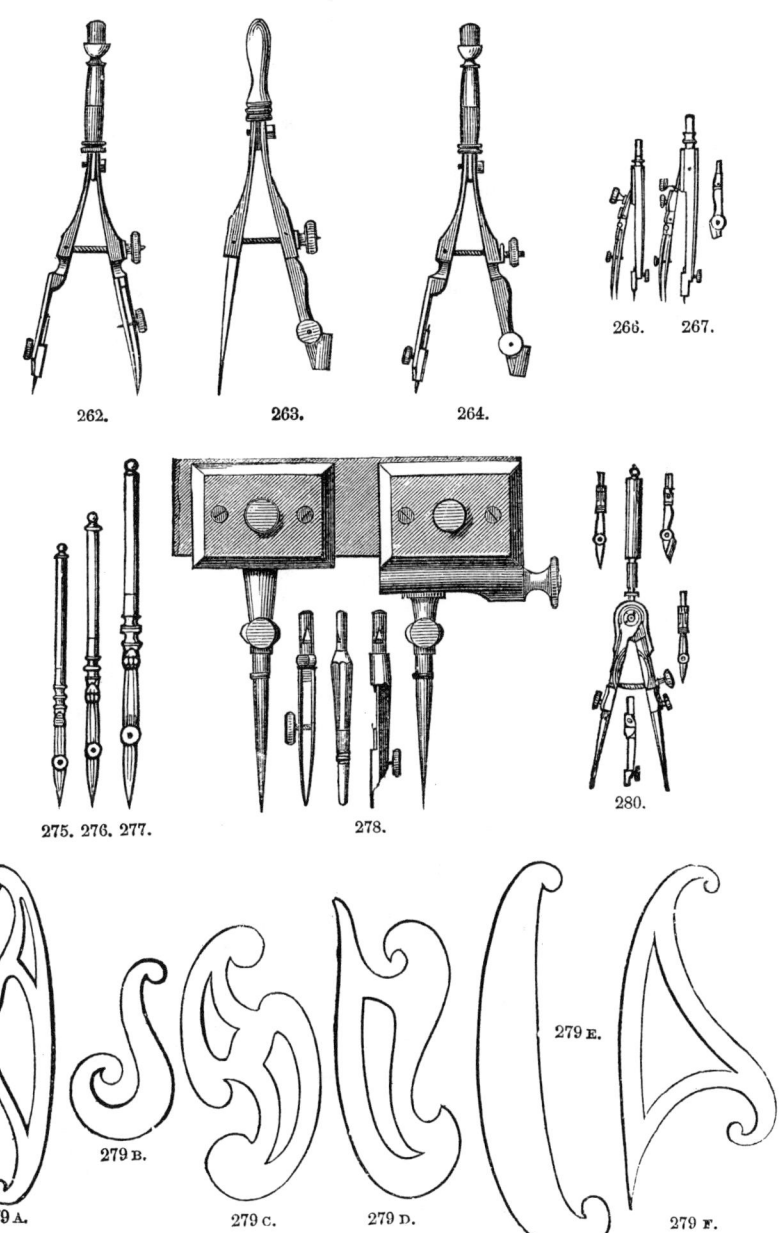

262.   263.   264.   266.   267.

275. 276. 277.   278.   280.

279 A.   279 B.   279 C.   279 D.   279 E.   279 F.

No.				Price.
275.—Drawing Pen, with joint, 4½ inches long				1 60
276.—	"	"	5½ "	1 70
277.—	"	"	6½ "	1 95
278.—Beam Compass furniture, for wood beams, $8 30; in Morocco box				8 75
279.—Horn Curves, A, B, C, D, E, F, each				75

280.—Drawing Compass, 4 inches, with long ivory handle, spring and micrometer, with 2 pens, pencil-holder and needle point........ .... 8 50

BOXES FOR DRAWING INSTRUMENTS.

☞ Parties wanting cases made up, can select the pieces, and we will
have boxes made to suit, at an additional cost of from $1 to $15, according to the size of the boxes, which are made of morocco, rosewood,
or walnut, highly finished.

SETS OF EXTRA FINE SWISS DRAWING INSTRUMENTS.

The following sets have beautifully finished Walnut Boxes, 9½ inches long
by 6 inches wide, with lock and key and tray.

300.—Contains pair plain Dividers, No. 231.
Set of Instruments, No. 226.
Steel Spacing Divider, No. 259.
Steel Bow-Pen, 3½ inches, No. 261.
Steel Bow-Pencil, 3½ inches, No. 263.
Drawing Pen, No. 276.
Triangular Scale, 6 inch $25 00
301.—Contains pair plain Dividers, No. 231.
Set of Instruments, No. 226.
Do.   No. 233.
Drawing Pen, No. 275.
Do.   No. 277.
Triangular Scale 6 inch... $26 50

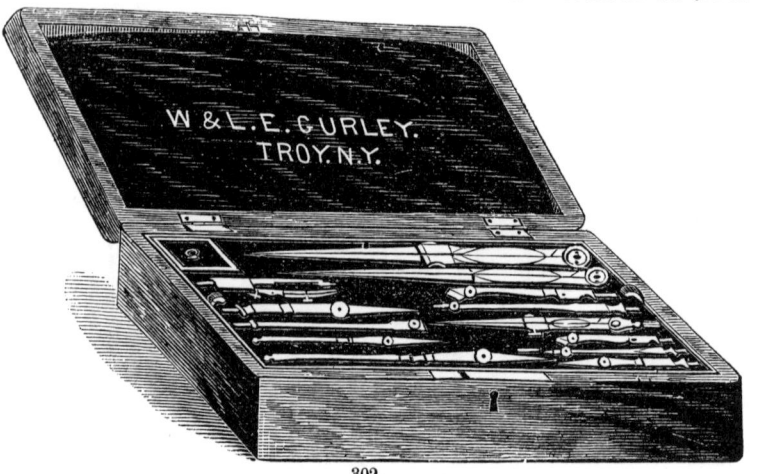

302.

No.   Price.

302.—Contains pair plain Dividers, No 231.
　　　Set of Instruments, No. 226.
　　　　　Do.　　　No. 233.
　　　Bow Pen, German Silver, No. 266.
　　　Drawing Pen, No. 275.
　　　　　Do.　　No. 277.
　　　Triangular Scale, 6 inch.. $29 00

303.—Contains pair plain Dividers, No. 231.
　　　Pair Hair Spring Dividers, No. 229.
　　　Set of Instruments, No. 226.
　　　Steel Spacing Dividers, No. 259.
　　　Steel Bow Pen, No. 261.
　　　Steel Bow Pencil, No. 263.
　　　Drawing Pen, No. 275.
　　　　　Do.　　No. 277.
　　　Triangular Scale, 6 inch $30 00

304.—Contains pair plain Dividers, No. 231.
　　　Pair Hair Spring Dividers, No. 229.
　　　Set of Instruments, Nos. 226 and 233.
　　　Bow Pen, German Silver, No. 266.
　　　Drawing Pens, Nos. 275 and 277.
　　　Triangular Scale, 6 inch $33 00

The following sets have beautifully finished Walnut Boxes, 13 inches long
by 6 inches wide, with lock and key and tray.

305.—Contains pair plain Dividers, No. 231.
　　　Set of Instruments, No. 226.
　　　Steel Bow Pen, No. 261.
　　　Drawing Pens, Nos. 275 and 276.
　　　Triangular Scale, 12 inch.. $23

306.—Contains pair plain Dividers, No. 231.
　　　Set of Instruments, Nos. 226 and 233,
　　　Steel Bow Pen, No. 261.
　　　Steel Bow Pencil, No. 263.
　　　Drawing Pens, Nos. 275 and 276.
　　　Triangular Scale, 12 inch $31 50

The following sets have beautifully finished Rosewood Boxes, 13 inches long
by 7½ inches wide, with lock and key and tray.

307.—Contains pair plain Dividers, No. 231.
　　　Pair Hair Spring Dividers, No. 229.
　　　Set of Instruments, Nos. 226 and 233.
　　　air Steel Spacing Dividers, No. 259.
　　　Steel Bow Pen, No. 261.
　　　Steel Bow Pencil, No.263.
　　　Drawing Pens, Nos. 275, 276 and 277.
　　　Triangular Scale, 12 inch........ .... $39 00

No.   Price.

308.—Contains pair plain Dividers, No. 231.
    Pair Hair Spring Dividers, No. 229.
    Set of Instruments, Nos. 226 and 23
    Proportional Dividers, No. 235.
    Steel Spacing Dividers, No. 259.
    Steel Bow Pen, No. 261.
    Steel Bow Pencil, No. 263.
    Drawing Pens, Nos. 275, 276 and 277.
    Triangular Scale, 12 inch.. $49 00

309.—Contains pair plain Dividers, No. 231.
    Pair Hair Springs Dividers, No. 229.
    Set of Instruments, Nos. 226 and 233.
    Proportional Dividers, No. 236.
    Steel Spacing Dividers, No. 259.
    Steel Bow Pen, No. 261.
    Steel Bow Pencil, No. 263.
    Beam Compass, No. 241.
    Drawing Pens, Nos. 275, 276 and 277
    Road Pen, No. 248.
    Dotting Pen, one wheel No. 245.
    Triangular Scale, 12 inch $70 00

The following set has beautifully finished Rosewood Box, $15\frac{1}{2}$ inches long by 10 inches wide, with lock and key and tray, and lined with finest silk velvet.

310.—Contains pair plain Dividers, No. 231.
    Pair Hair Spring Dividers. No. 229.
    Set of Instruments, No. 225.
    Proportional Dividers, No. 238.
    Steel Spacing Dividers, Nos. 258 and 259.
    Beam Compass, No. 242.
    Steel Bow Pen, No. 261.
    Set of Instruments, No. 280.
    Steel Bow Pencil, No. 263.
    Drawing Pens, Nos. 275, 276 and 277.
    Road Pen, No. 247.
    Dotting Pen with 6 wheels, No. 245.
    Protractor.
    Triangular Scale, 12 inch
    Set of Color Cups,.. $105 00

ALTENEDER'S PATENT JOINT DRAWING INSTRUMENTS.

The excellency of these instruments consists in the joints of the dividers being so constructed as to prevent any irregular motion when the legs are opened or closed, also for the general care with which the instruments are finished.

All the pens are thoroughly well made and pointed. No. 315 represents a sectional view of Alteneder's Patent Joint Divider Head.

☞ Parties wanting cases made up, can select the pieces, and we will have boxes made to suit, at an additional cost of from $3 to $15, according to the size of the boxes, which are made of morocco, rosewood, or walnut, highly finished.

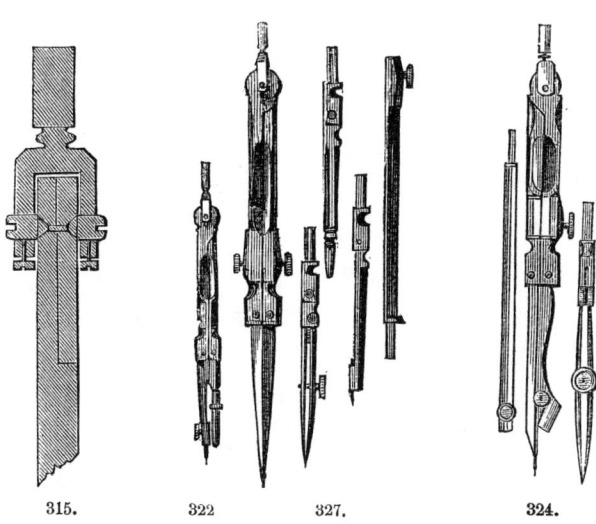

315.   322   327.   324.

No.   Price.

316.—Plain Dividers of German Silver, 3½ inches long, with Alteneder's patent joint, each $2 00

317.—Plain Dividers of German Silver, 5 inches long, with Alteneder's patent joint, each... 2 75

318.—Plain Dividers of German Silver, 6 inches long, with Alteneder's patent joint, each.. 3 25

319.—Hair Spring Dividers of German Silver, 3½ inches long, with Alteneder's patent joint, each... 3 00

320.—Hair Spring Dividers of German Silver, 5 inches long, with Alteneder's patent joint, each... 3 50

321.—Hair Spring Dividers of German Silver, 6 inches long, with Alteneder's patent joint, each... 4 00

322.—Needle Point Dividers, 3½ inches long, of German Silver, with Pencil Point and Alteneder's patent joint, each..................................... 4 25

323 —Needle Point Dividers, 3½ inches long, of German Silver, with pen Point and Alteneder's patent joint, each...................................... 4 75

324.—Needle Point Dividers, 6 inches long, of German Silver, with Pen and Pencil Point and Lengthening Bar, and Alteneder's patent joint....... 7 50

325.—Needle Point Dividers, 3½ inches long, of German Silver, with Pen and Pencil Point, and Alteneder's patent joint............................ 6 25

326.—Steel Point Dividers, 6 inches long, with Pen, Pencil, Needle Point and Lengthening Bar.. 9 00

No.						Price.
327.—Steel Point Dividers, 6 inches long, with Pen, Pencil, Needle Point, Lengthening Bar and joint in each leg						$12 00
328.—Steel Point Dividers, 3½ inches long, with Pen, Pencil, and Needle Point.						6 50
329.—Steel Spacing Dividers, 3 inches long						2 00
330 " " " Metal Handle						2 00
331.—Steel Bow Pen, 3 inches long, round points						2 50
332.— " " " with Needle Point						3 25
333,— " " " " " Metal Handle						3 25
334.— " Pencil " with round point						2 50
335.— " " " with Needle Point						3 25
336.— " " " " " Metal Handle						3 25
337.—Drawing Pens, 4½ inches long						1 60
338.— " 5½ "						1 70
339.— " 6½ "						1 95

BRASS DRAWING INSTRUMENTS,

FOR SCHOOLS.

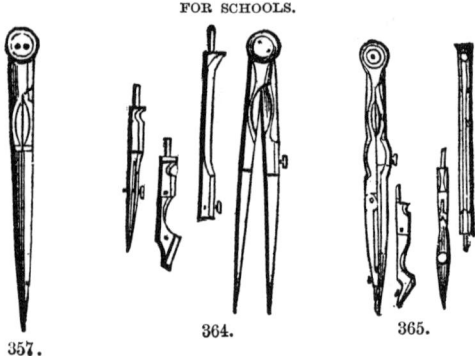

357.   364.   365.

350.—Wood Dividers, 13 in. long, with crayon holder, for black-board drawing.						$1 00
351.— " 16 " " " "						1 25
352.— " 20 " " " "						1 50
353.— " 24 " " " "						1 75
354.— " 27 " " " "						2 00
355.— " 30 " " " "						2 25
356.— " 36 " " " "						2 50
357.—Brass Dividers, 3½ inches long, screw joint						30
358.— " 4½ " "						35
359.— " 5½ " "						45
360.— " 6½ " "						60
361.— " 4½ " rivet joint						25
362.— " 5½ " "						35
363.— " 6½ " "						45
364.—Brass Dividers, 4½ inches long, with Pen, and Pencil Points and Lengthening Bar						85
365.—Brass Dividers, 6 inches long, with Pen and Pencil Points and Lengthening Bar						1 00

No.   Price.

366.—Brass Dividers, Needle Point, 4½ inches long, with Pen and Pencil Points
    and Lengthening Bar.. $1 00

367.—Brass Dividers, Needle Point, 6 inches long, with Pen and Pencil Points
    and Lengthening Bar.. 1 C0

370.—Dividers, brass, medium quality, needle point, with pen and pencil
    points, 3 inches 75

371.—Bow Pencil, brass... 75

372.—Bow Pen, brass, needle points, no spring................................. 75

373.—Bow Pen, brass, needle points, and adjusting spring..................... 75

374.—Furniture for Beam Compass, brass, with adjusting screw, in morocco
    case 5 50

375.—Bisecting Dividers, brass 75

376.—Proportional Dividers, brass, divided for lines, in case 2 50

277.—Drawing Pen, black handle.................................... 25

No.		Price.
378.—Drawing Pen, ivory handle	....	$ 40
379.—Roulette for dotting lines, with extra wheels.	....	1 00
380.—Double Drawing or Road pen, Brass mounted	....	2 25

CASES OF BRASS DRAWING INSTRUMENTS

FOR SCHOOLS.

385.

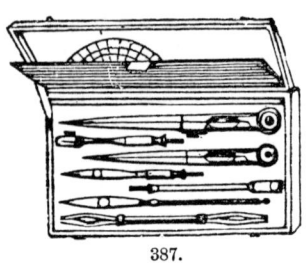

387.

385.—Wood Box ; pair 4½-inch Dividers, with pen and pencil points, and
Crayon Holder... $ 75

386.—Wood Box ; pair 4½-inch Dividers, with pen and pencil points and lengthening bar ; Ebony handle Drawing Pen ; Boxwood scale, 4 inches
long.. 1 25

387.—Wood Box ; Pair of 4½-inch Dividers, with pen and pencil points and
lengthening bar ; Pair of 3½-inch plain Dividers, Drawing Pen, Horn
Protractor ; Boxwood Scale, 4 inches long 1 35

388.—Rosewood Box ; Pair 5½-inch Dividers, with pen and pencil points and
lengthening bar ; Pair of 4½-inch plain Dividers, Drawing Pen, Horn
Protractor ; Box Wood Scale, 6 inches long........................... 1 85

389.—Same as 388, with addition of Parallel Ruler 2 10

390.—Same as 388, with Ivory Scale, 6 inches long............................. 2 50

391.—Same as 390, with addition of Parallel Ruler............................. 2 75

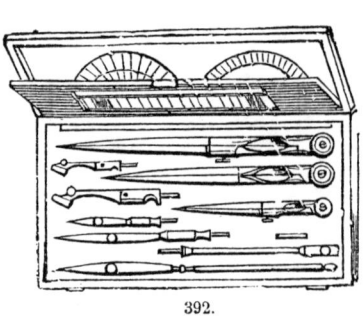

392.

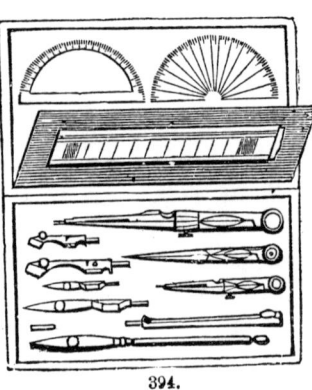

394.

No.   Price.

392.—Rosewood Box ; Pair of 6-inch Dividers, with pen and pencil points and lengthening bar Pair of 4½-inch plain Dividers, Drawing Pen ; Pair of 3½-inch Dividers, with pen and pencil points ; Brass Protractor, Horn Protractor ; Ivory Scale, 6 inches long, per set $3 35

393.—Same as No. 392, but with the instruments set in a tray, so that colors, etc., may be put below, per set.................................... 3 85

394.—Rosewood Box ; Pair of 6-inch needle point Dividers, with pen and pencil points, and lengthening bar ; Pair 4½-inch plain Dividers ; Pair of 3½-inch needle point Dividers, with pen and pencil points ; Drawing Pen, Brass Protractor, Horn Protractor ; Ivory Scale, 6 inches long, per set· 4 00

395.—Same as No. 394, but with lock and key and the instruments set in a tray, so that the colors may be put bel ow, per set........................... 4 50

396.—Same as No. 395, with addition of Parallel Ruler, per set................ 4 75

397.—Same as No. 396, with patent Pencil holder, to the 6-inch and 3½ inch Dividers.. 5 25

398.—Rosewood Box, with lock and key, the instruments set in a tray, so that colors, etc., may be put below ; Pair of 6-inch needle point Dividers, with pen and pencil points, and lengthening bar ; Drawing Pen, pair 4½-inch plain Dividers, Brass Protractor, Horn Protractor, Pair of 3½-inch Needle Point Dividers, with pen and pencil points ; Spring Bow Pen, with needle point ; Ivory Scale, 6 inches long, per set 5 35

399.—Same as No. 398, with addition of Parallel Ruler, per set................ 5 60

400.—Same as No. 398, with patent Pencil holder, to the 6-inch and 3½ inch Dividers.. 6 00

401.—Same as No. 400, with the addition of a pair Proportional Dividers, has no brass Protractor, but has wood Triangle and Irregular Curves...... 7 25

402.—Same as No. 401 with patent pencil holder, to the 6-inch and 4½ inch Dividers 8 00

FINE GERMAN SILVER INSTRUMENTS.

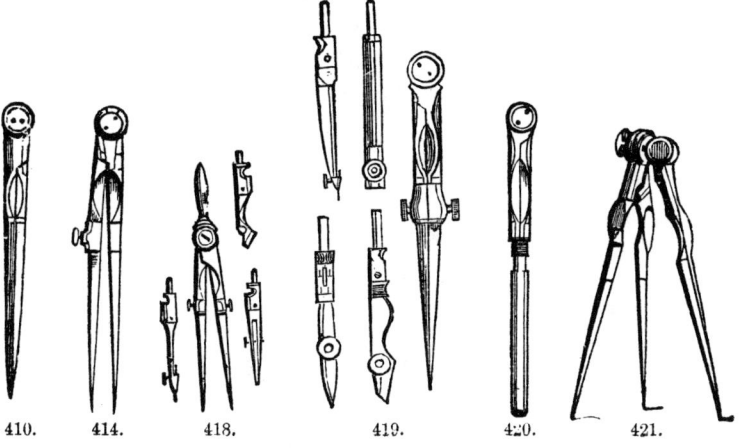

410.   414.   418.   419.   420.   421.

No.							Price.
410.—Dividers, German Silver, steel joints, turned cheeks, fine finish, 4 inch..							$ 70
411.—	"	"	"	"	"	5 " ..	80
412.—	"	"	"	"	"	6 " ..	1 00
413.—	"	"	"	"	"	7 " ..	1 25

414.—Hair Spring Dividers, German Silver; steel joints, turned cheeks, fine finish, 4 inch.. 1 25

415.—Hair Spring Dividers, German Silver; steel joints, turned cheeks, fine finish, 5 inch.. 1 80

416.—Hair Spring Dividers, German Silver; steel joints, turned cheeks, fine finish, 6 inch.. 2 50

417.—German Silver Plain Dividers, 3 inches long, with Handle.............. 1 60

418.—Dividers, German Silver; fine quality, needle point, with pen and pencil point, 3 inches.. 2 75

419.—Dividers, German Silver; fine quality, with needle point, pen, lengthening bar, and pencil points, 6 inches.................................... 3 50

420.—Dividers, German Silver; 5-inch, fine finish, with shield for pocket..... 2 00

421.—Dividers, German Silver; 5-inch, three-legged.......................... 3 50

425.   427.   428.   429.   430.   433. 434.   437.
                                       431. 432.

425.—Proportional Dividers, German Silver, 6½ inches long, divided for lines.. 2 50

426.—Proportional Dividers, German Silver, 6½ inches long, divided for lines, circles. plans, and solids... 3 60

427.—Bisecting Dividers, German Silver............................... ... 1 12

428.—Spacing Dividers, all steel, with Spring and Adjusting Screw............ 1 25

429.—Bow Pen, all steel, ivory handle...................................... 1 50

430.—Bow Pencil, all steel, ivory handle.................................... 1 50

431.—Spring Bow Pen, German Silver.......... .... 1 62

432.—   "   "   with pencil point...................... 2 50

No.   Price.

433.—Drawing Pen, German Silver, medium finish, hinge to pen.............. $ 60

434.—Drawing pen, German Silver, fine finish, hinge to pen...................   75

435.—Drawing Pen, German Silver, fine finish, hinge to pen, and protracting

      pin.. 1 00

436.--Drawing Pen, all German Silver, for red ink.............................   75

437.—Double Drawing, or Road Pen 2 75

438.—Triple   "   for three lines at one time......................... 4 50

439.—Dotting Pen, New Style, two wheels, in case............................ 3 50

440.—Dotting Pen, one wheel 2 65

441.—   "   six wheels... 4 00

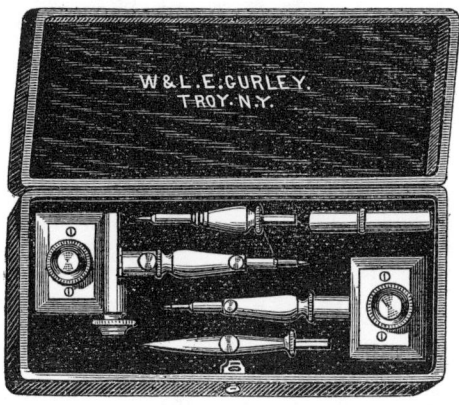

442.

442.—Furniture for Beam Compasses, German Silver, with adjusting screw, in

      morocco case... .... 6 50

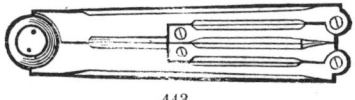

444.

443.—Pocket Dividers, German Silver, folding pen and pencil points.......... 5 00

444.—Map perambulator for measuring the length of curved lines, rivers, rail-

      roads, etc. on maps, each................. .... 1 50

    ☞ Parties wanting cases made up, can select the pieces, and we will
have boxes made to suit, at an additional cost of from $1 to $15, according to the size of the boxes, which are made of morocco, rosewood, or
walnut, highly finished.

CASES OF FINE GERMAN SILVER INSTRUMENTS.

FOR ENGINEERS, ARCHITECTS, AND MACHINISTS.

No.   Price.

450.—Morocco Box; pair of 5½-inch Dividers, with Pen and Pencil Points, Drawing Pen, Ivory Scale, 6 inches long, same as in school-cases of instruments.. $3 50

451.—Morocco Box; pair of 3-inch Dividers, with Pen, Pencil, and Needle Points, and Lengthening Bar, Drawing Pen. No Scale or Protractor. Per set. 5 00

451.   452.

452.—Morocco Box; pair of 5½-inch Dividers, with Pen and Pencil Points, pair of 5-inch plain Dividers, Drawing Pen, Ivory Protractor Scale, 6 inches long, per set... 5 00

453.— Same as No. 452, with addition of Needle Points and Lengthening Bar, to to 5½-inch Dividers, per set.. . 6 50

454.—Morocco Box, rounded corners, for carrying in the pocket; pair of 4¾-inch Dividers, with hinge in one leg, Needle Points, with Pen and Pencil Points, and Lengthening Bar, Spring Bow Pen, Needle Point, pair of 4-inch Plain Dividers, rounded point, Drawing Pen, ivory handle, 5-inch Ivory Rule, divided into eighths, per set.................. 8 50

455.   456.

455.—Morocco Box; pair of 5½-inch Dividers, with Pen and Pencil Points, and Lengthening Bar, pair of 5-inch plain Dividers, pair of 3-inch Dividers, with Pen and Pencil Points, Drawing Pen, German Silver Protractor, German Silver Square, Ivory Scale, 6 inches long, per set............ . 9 50

456.—Morocco Box; pair of 5½-inch Dividers, with Pen, Pencil and Needle Points, and Lengthening Bar, pair of 5-inch Plain Dividers, Spring Bow Pen, Drawing Pen, Ivory Protractor Scale, 6 inches long, per set. 9 50

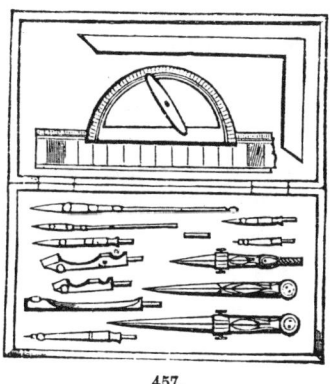

457.

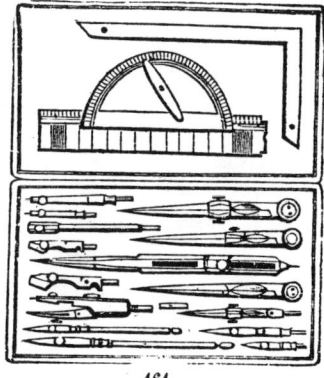

464.

No.   Price.

457.—Morocco Box ; pair of 5½-inch Dividers, with Pen, pencil and Needle
    Points, and Lengthening Bar, pair of 5-inch plain Dividers, pair of
    3-inch Dividers, with Pen, Pencil and Needle Point, 2 Drawing Pens,
    German Silver Square, German Silver Protractor, Ivory Scale, 6 inches
    long, per set $11 00

458.—Same instruments as in No. 457, in polished Walnut Box, per set....... 14 50

459.

No.   Price.

459.—" R. P. I." Polished Walnut Box, with lock and key and tray, containing pair 5½ inch Dividers, Pen, Pencil and Needle Point, 1 pair 5 inch Hair Spring Dividers, pair 3 inch Dividers, Pen, Pencil and Needle Point, 2 Swiss Pens, Nos. 275 and 277.............................. $15 00

460.—Polished Walnut Box ; containing pair 5½ inch Dividers, with Pen, Pencil and Needle Points, and Lengthening Bar,

       Pair 5 inch plain Dividers,

       Pair of 3 inch Dividers, with Pen, Pencil and Needle Points,

       Spring Bow Pen, with Needle Point,

       2 Drawing, Pens,

       German Silver Square,

       German Silver Protractor,

       Ivory Scale, 6 inches long,................................... .... 15 50

461.—Same as No. 460, in Polished Walnut Box, with lock and key and tray,.. 17 50

462.—Polished Walnut Box ; containing pair 5½ inch Dividers, with Pen Pencil and Needle Points and Lengthening Bar,

       Pair of 5 inch plain Dividers,

       Pair of 5 inch Hair Spring Dividers,

       Pair of 3 inch Dividers, with Pen, Pencil and Needle Points,

       Spring Bow Pen, with Needle Point,

       2 Drawing Pens,

       German Silver Square,

       German Silver Protractor,

       Ivory Scale, 6 inches long, ... 17 25

463.—Same as No. 462, set in a tray, and the box much larger, with lock, thus affording space for extra instruments or colors, &c., ... 18 75

464.—Polished Walnut Box, with lock and key·and tray ; containing pair 6 inch Dividers, with Pen, Pencil and Pen Point, and Lengthening Bar,

       Pair 5 inch plain Dividers,

       Pair 5 inch Hair Spring Dividers,

       Pair 3 inch Dividers, with Pen, Pencil and Needle Point,

       Bow Pen, German Silver,

       2 Drawing Pens,

       1 Red Ink Pen, 1 Road Pen,

       Pair Proportional Dividers, No. 425.

       Protractor, 4 inch, half circle, whole Degrees.

       Triangle, and Triangular Scale, 12 inch.............................. 27 00

465.—Same as No. 464, with addition of Beam Compass........................ 32 00

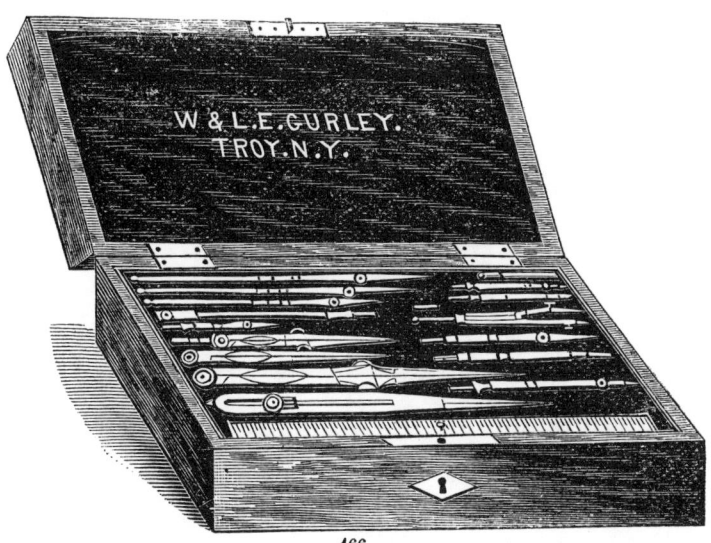

466.

466.—Polished Rosewood Box, inlaid, lock and key, with tray, leaving space
    below for paints, rules &c., containing pair of 6½ inch Needle Point
    Dividers, with Pen and Pencil Points and Lengthening Bar.
    Pair 4½ inch plain Dividers,
    Pair of 4 inch, Needle Point Dividers, with Pen and Pencil Points.
    Pair of Proportional Dividers. No. 426.
    3 Drawing Pens,
    Horn Protractor,
    1 Wood Curve and 2 Wood Squares,
    Bow Pen, German Silver,
    Ivory Rule, 8 inches long,
    Ivory Scale, 6 incnes long,... $27 50
467.—Same as No. 466, with Patent Pencil Points, to the 6 inch and 4 inch
    Dividers 28 50
468.—Polished Rosewood Box, inlaid, with brass edges, lock and key, with tray
    leaving space below for paints, rules, &c. ; containing pair of 6 inch
    Needle Point Dividers, with Pen and Pencil Points and Lengthening Bar.
    Pair 4½ inch plain-Dividers,
    Pair of 4 inch Dividers, Needle Points, with Pen and Pencil Points.
    Pair of Proportional Dividers. No. 435.
    Bow Pen German Silver,
    3 Drawing Pens.
    Furniture for Beam Compass, with Micrometer Screw.
    9 inch Horn Protractor.
    Ivory Scale,6 inches long,
    Ivory Scale, 8 inches long, one edge divided to inches and eighths, the
    other to centimeters and millimeters,............................. 34 00
469.—Same as No. 468, with Patent Pencil Points, to the 6 in. and 4 in. Dividers 35 00

EXTRA FINE SWISS PROTRACTORS.

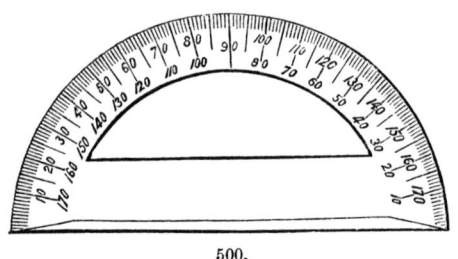

500.

No.								Price.
500.—Protractor, 4 in. diameter, ½ circle, whole degrees, centre on outer edges.								$1 90
501.—	Do.	5	do.	½ do.	½	do.	do.	do.	2 50
502.—	Do.	6	do.	½ do.	½	do.	do.	do.	3 25
503.—	Do.	6	do.	½ do.	¼	do.	do.	do.	4 00

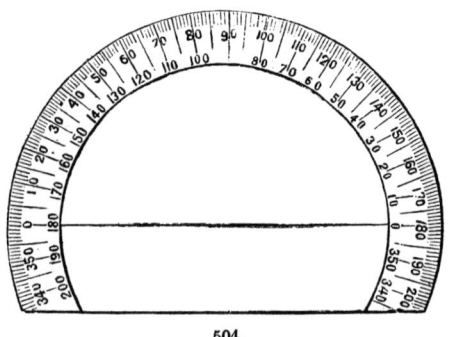

504.

504.—Protractor, 5 in. diameter, ½ circle, ½ degree, centre on inner edge								.... 2 50
505.—	Do.	6	do.	½ do.	½	do.	do.	do.	.... 3 50
506.—	Do.	6	do.	½ do.	¼	do.	do.	do.	.... 4 50

EXTRA FINE SWISS PROTRACTORS, OF GERMAN SILVER, WITH ARMS.

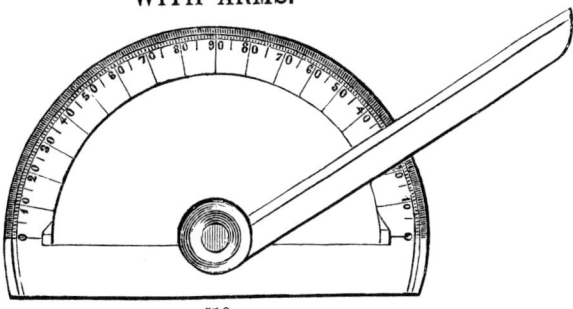

510.

No.   Price·

510.—German Silver Protractor, 5 inches diameter, half circle, with Arm, and
      divided in half degrees, ... $6 50
511.—German Silver Protractor, 6 inches diameter, half circle, with Arm, and
      divided in half degrees, ... 7 50
512.—German Silver Protractor, 7 inches diameter, half circle, with Arm, and
      divided in half degrees,............................. .... 9 00
513.—German Silver Protractor, 8 inches diameter, half circle, with Arm, and
      divided in half degrees,.. 11 00
514.—German Silver Protractor, 5 inches diameter, whole circle, with Arm and
      divided in half degrees,.......................... .... 9 50
515.—German Silver Protractor, 6 inches diameter, whole circle, with Arm, and
      divided in half degrees,.. 11 50
516.—German Silver Protractor, 7 inches diameter, whole circle, with Arm, and
      divided in half degrees,......................... .... 13 00
517.—German Silver Protractor, 8 inches diameter, whole circle, with Arm, and
      divided in half degrees, ... 15 00

EXTRA FINE SWISS PROTRACTORS, OF GERMAN SILVER, WITH ARMS AND VERNIERS.

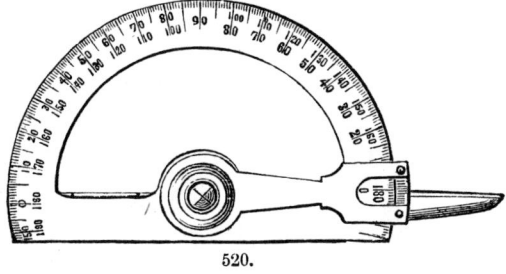

520.

520.—Protractor, 5½ inches diameter, half circle, half degrees, with vernier
      reading to three minutes..................................... $11 00

No.   Price.

521.—Protractor, 8 inches diameter, half circle, quarter degrees, with vernier
reading to one minute..$ 14 50

522.—Protractor, 10 inches diameter, half circle, quarter degrees, with vernier
reading to one minute.. 18 00

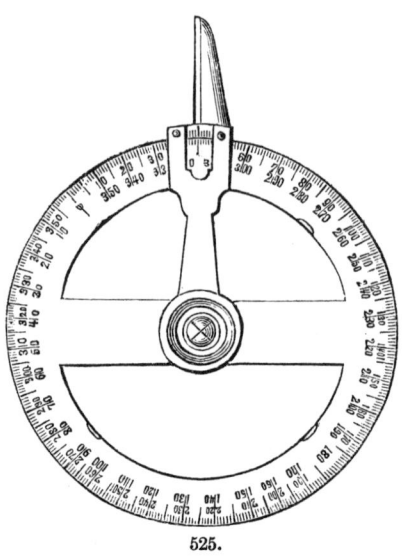

525.

525.—Protractor, 5½ inches diameter, whole circle, half degrees, with vernier
reading to three minutes... 14 00

526.—Protractor, 8 inches diameter, whole circle, quarter degrees, with vernier reading to one minute... 16 00

527.—Protractor; 10 inches diameter, whole circle, quarter degrees, with vernier reading to one minute... 20 00

☞Cases for Protractors, of wood, covered with morocco, lined with velvet
according to size: from $1 00 to $5 00.

PROTRACTORS OF HORN, BRASS, GERMAN SILVER, IVORY
AND PAPER.

535.—Railroad Curve Protractor, of horn, 8 inches diameter, having laid off on
it twenty-three curves from ½ degree to 8 degrees, with a radius of 400
feet to the inch...$ 2 00

536.—Horn Protractors, 5 inches diameter, whole circle, half degrees.......... 1 00

537.—   "   6   "   "   " 1 50

538.—   "   7   "   "   " 1 75

539.—Horn Protractors, 4 inches diameter, half circle, whole degrees.......... 20

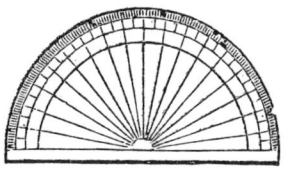

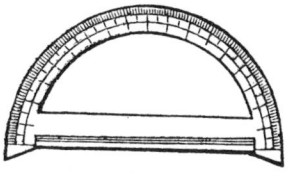

540.   545.

No.							Price.
540.—Horn Protractors, 5 inches diameter, half circle, half degrees............							$ 35
541.—	"	6	"	"	"	....	45
542.—	"	7	"	"	"	....	75
543.—	"	8	"	"	"	....	1 00
544.—Brass Protractor, 4			"	"	whole degrees..........		20
545.—	"	4	"	"	half degrees............		40
546.—	"	5	"	"	"	....	60
547.—	"	6	"	"	"	....	75
548.—German Silver Protractor, 4 inches diameter, half circle, whole degrees..							50
549.—	"	"	5	"	"	half degrees....	1 00
550.—	"	"	6	"	"	"	....	1 25
551.—	"	"	7	"	"	"	....	1 50
552.—	"	"	5	"	" beveled edge, half deg.		1 50
553.—	"	"	6	"	"	"	"	2 00
554.—	"	"	7	"	"	"	"	3 00

PAPER PROTRACTORS.

560.—Whole Circle Protractor, 13 inches diameter, half degrees, on drawing
   paper, each... $ 30

561.—Whole Circle Protractor, 13 inches diameter, half degrees, on Bristol
   boards, each... 40

562.—Half Circle Protractor, whole degrees, 4 inch, Bristol boards............ 20

563.—Half Circle Protractor, 6 inches diameter, half degrees, on Bristol boards,
   each 25

564.—Half Circle Protractor, 8 inches diameter, half degrees, on Bristol boards,
   each.. 30

565.—Circular Protractor on tracing paper, 14 inches diameter, quarter degrees
   (these are used by the U. S. Coast Survey, and U. S. Navy, and give
   entire satisfaction.)... 30

IVORY PROTRACTORS.

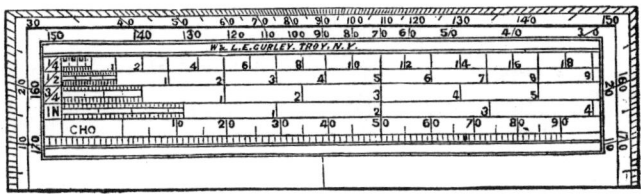

575. FRONT SIDE.

No.		Price.
575.	—Ivory Rectangular Protractor, 6 inches long, $1\frac{3}{4}$ inches wide, with scales as follows: front sides divided around edges from 0 to 180 degrees in single degrees, scales of $\frac{1}{4}$, $\frac{1}{2}$, $\frac{3}{4}$ and 1 inch to the foot. and scale of chords. Reverse side scales of 30, 35, 40, 45, 50 and 60 parts to the inch, scale of chords and diagonal scale of inches and $\frac{1}{100}$ths..........	$1 75
576.	—Ivory Rectangular Protractor, 6 inches long by $1\frac{3}{4}$ inches wide, with scales as follows: front side, the edge divided in single degrees from 0 to 180 degrees, scales of $\frac{1}{8}$, $\frac{1}{4}$, $\frac{3}{8}$, $\frac{1}{2}$, $\frac{5}{8}$, $\frac{3}{4}$, $\frac{7}{8}$, and 1 inch to the foot, and scale of chords. On the reverse sides, scales of 30, 35, 40, 45, 50 and 60 parts to the inch. scale of chords and diagonal scale of $\frac{1}{100}$ths...	3 00
577.	—Ivory Rectangular Protractor, 6 inches long by 2 inches wide, with scales as follows: front side, the edge divided in single degrees from 0 to 180 degrees, scales of $\frac{1}{8}$, $\frac{1}{4}$, $\frac{3}{8}$, $\frac{1}{2}$, $\frac{5}{8}$, $\frac{3}{4}$, $\frac{7}{8}$, 1, $1\frac{1}{8}$, $1\frac{1}{4}$ inches to the foot, scale of chords, and line of 40 parts on lower edge. On the reverse side, scales of 20, 25, 30, 35, 40, 45, 50, 60 parts to the inch, diagonal scale of $\frac{1}{100}$ths..	3 25
578.	—Ivory Rectangular Protractor, same as No. 577, but has the Protractor divided in $\frac{1}{2}$ degrees....	4 00
579.	—Ivory Rectangular Protractor, 6 inches long by $2\frac{1}{4}$ inches wide, with scales as follows: front side, the edge divided in $\frac{1}{2}$ degrees from 0 to 180 degrees, scales of $\frac{1}{8}$, $\frac{1}{4}$, $\frac{3}{8}$, $\frac{1}{2}$, $\frac{5}{8}$, $\frac{3}{4}$, $\frac{7}{8}$, 1, $1\frac{1}{8}$, $1\frac{1}{4}$, $1\frac{3}{8}$, $1\frac{1}{2}$ inches to the foot, scale of chords, and scale of 40 parts on lower edge. Reverse side, scales of 10, 15, 20, 25, 30, 35, 40, 45, 50, 60 parts to the inch, and diagonal scale of $\frac{1}{100}$ths..	4 50
580.	—Ivory Rectangular Protractor, 6 inches long by $2\frac{1}{2}$ inches wide, with scales as follows: front side, the edge divided in $\frac{1}{2}$ degrees from 0 to 180 degrees, scales of $\frac{1}{8}$, $\frac{1}{4}$, $\frac{3}{8}$, $\frac{1}{2}$, $\frac{5}{8}$ $\frac{3}{4}$, $\frac{7}{8}$, 1, $1\frac{1}{8}$, $1\frac{1}{4}$, $1\frac{3}{8}$, $1\frac{1}{2}$ inches to the foot, scale of chords, and scale of 40 parts on lower edge. Reverse side, scales of 20, 25, 30, 35, 40, 45, 50 and 60 parts to the inch, 2 scales of chords, scales of latitude, sines, tangents, hours, longitudes, secants, rhombs..	6 00
581.	—Ivory Rectangular Protractor, 8 inches long by 2 inches wide, with scales as follows: front side, the edge divided in $\frac{1}{2}$ degrees from 0 to 180 degrees, scales of $\frac{1}{8}$, $\frac{1}{4}$, $\frac{3}{8}$, $\frac{1}{2}$, $\frac{5}{8}$, $\frac{3}{4}$, $\frac{7}{8}$, 1 inch to the foot, scale of chords and scale of 40 parts on lower edge. Reverse side, scales of 30, 35, 40, 45, 50, 60 parts to the inch, scale of chords and diagonal scale of $\frac{1}{100}$ths.	5 00
582.	—Ivory Rectangular Protractor, 12 inches long by $2\frac{1}{2}$ inches wide, with scales as follows: the edge divided in $\frac{1}{2}$ degrees from 0 to 180 degrees, scales of $\frac{1}{8}$, $\frac{1}{4}$. $\frac{3}{8}$, $\frac{1}{2}$, $\frac{5}{8}$, $\frac{3}{4}$, $\frac{7}{8}$, 1, $1\frac{1}{8}$, $1\frac{1}{4}$, $1\frac{3}{8}$, $1\frac{1}{2}$, scale of chords and scale of 40 on lower edge. Reverse side, scales of 10, 15, 20, 25, 30, 35, 40, 45, 50, 60 parts to the inch, scale of chords and diagonal scale of $\frac{1}{100}$ths............................	11 50

IVORY SECTORS AND SCALES.

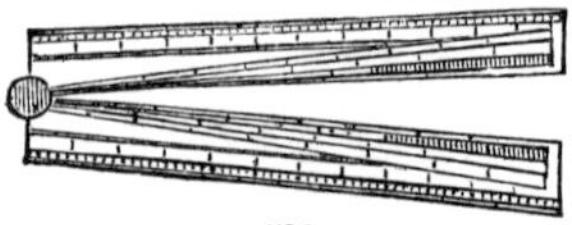

590,

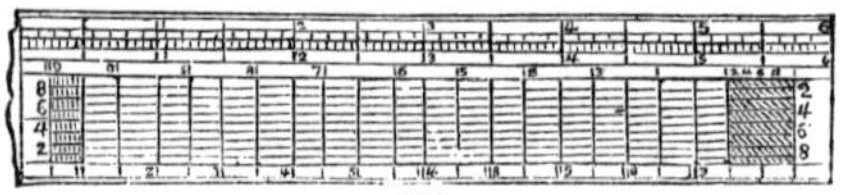

591.

590.—Ivory Sector, 6 inches long, opens to 12 inches long...................... $2 25

591.—Ivory Scale, 6 inches long, for school drawing.......................... 75

IVORY CHAIN SCALES.

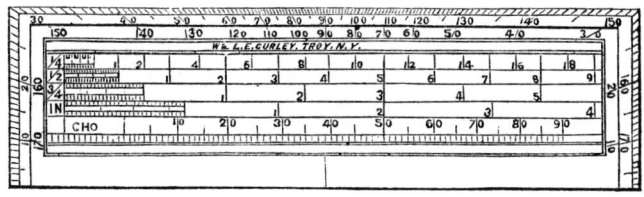

593.

No.   Price.

593.—Ivory Chain Scales, 12 inches long, graduated on two edges with either 10 and 10 parts, or 10 and 20, or 20 and 40, or 30 and 50, or 40 and 60, or 50 and 60, each.. $3 00

594.— Do.   do.   do.   with 40 and 80, or 50 and 100, each... 5 25

595.— Do.   do.   do.   with 80 and 100, each................. 5 75

596.—Ivory Off-set Scales, 2 inches long, 10 by 10, 10 by 20, 20 by 40, 30 by 50, 40 by 60, each.. 60

IVORY ARCHITECTS' SCALES.

597.

597.—Ivory Scale, 12 inches long, with 16 scales, as follows: $\frac{1}{8}$, $\frac{3}{16}$, $\frac{1}{4}$, $\frac{3}{8}$, $\frac{1}{2}$, $\frac{5}{8}$, $\frac{3}{4}$, $\frac{7}{8}$, 1, 1$\frac{1}{4}$, 1$\frac{1}{2}$, 1$\frac{3}{4}$, 2, 2$\frac{1}{4}$, 2$\frac{1}{2}$ and 3 inches to the foot, the first division of each scale subdivided in 12 parts, each.................... $3 00

598.—Same as No. 597, but with the first division of each scale subdivided into 10 parts, each... 3 00

599.—Ivory Scale, 12 inches long, with 12 scales, as follows: $\frac{1}{8}$, $\frac{3}{16}$, $\frac{1}{4}$, $\frac{3}{8}$, $\frac{5}{8}$, $\frac{7}{8}$, 1, 1$\frac{1}{4}$, 1$\frac{1}{2}$, 1$\frac{3}{4}$, 2 and 3 inches to the foot, the first division of each scale subdivided into 12 parts, diagonal scale reading to $\frac{1}{100}$ and $\frac{2}{100}$ of an inch, each....................... .. .. 3 00

600.—Same as No. 599, but has the first division of each scale subdivided into 10 parts, each... 3 00

601.—Ivory Scale, 12 inches long, one side rounded, the other flat, with the following scales, the graduations of which are all brought to the edge: $\frac{1}{16}$, $\frac{1}{8}$, $\frac{3}{16}$, $\frac{1}{4}$, $\frac{3}{8}$, $\frac{1}{2}$, $\frac{5}{8}$, $\frac{3}{4}$, $\frac{7}{8}$, 1, 1$\frac{1}{4}$, 1$\frac{1}{2}$, 1$\frac{3}{4}$, 2, 2$\frac{1}{2}$ and 3 inches to the foot, the first division of each scale is subdivided into twelve parts, each.. 3 00

602.—Same as No. 601, but the first division of each scale subdivided into ten parts, each... 30

BOXWOOD SCALES AND PROTRACTORS.

605. FRONT SIDE.

No.   Price.

605.—Boxwood Protractor, 6 inches long, 1¾ inches wide, whole degrees, with
6 scales of equal parts, 4 scales of feet and inches, 2 scales of chords,
and diagonal scale $0 50

606.—Boxwood Scale, 6 inches long, same as in School Cases of Instruments..   20

BOXWOOD CHAIN SCALES.

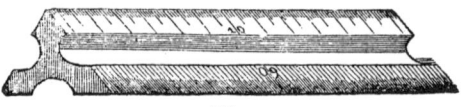

608.

608.—Boxwood Chain Scale, 12 inches long, graduated on two edges with either
10 and 10 parts, or with 10 and 20 parts, or with 20 and 40 parts, or with
30 and 50 parts, or with 40 and 60 parts, or with 50 and 60 parts........ $1 25

609.—Boxwood Off-set Scales, 2 inches long, graduated 10 by 10, 10 by 20, 20 by
40, 30 by 50, 40 by 60, each 25

ARCHITECTS' BOXWOOD SCALES.

610.

610.—Boxwood Scale, 12 inches long, with 16 scales, as follows: $\frac{1}{8}$, $\frac{3}{16}$, $\frac{1}{4}$, $\frac{3}{8}$,
$\frac{1}{2}$, $\frac{5}{8}$, $\frac{3}{4}$, $\frac{7}{8}$, 1, 1¼, 1 ½, 1¾, 2, 2¼, 2½ and 3 inches to the foot, the
first division of each scale subdivided in 12 parts, each $1 25

611.—Same as No. 610, but with the first division of each scale subdivided into
ten parts, each..   ....   1 25

612.—Boxwood Scale, 12 inches long, with 12 scales, as follows; $\frac{1}{8}$, $\frac{3}{16}$, $\frac{1}{4}$, $\frac{3}{8}$,
$\frac{5}{8}$, $\frac{7}{8}$, 1, 1¼, 1½, 1¾, 2 and 3 inches to the foot, the first division of
scale subdivided into 12 parts, and diagonal scale reading to $\frac{1}{100}$ths and
$\frac{1}{200}$ths of an inch, each............................   ....   1 25

613.—Same as No. 612, but has the first division of each scale subdivided into
10 parts, each.... ....   1 25

614.—Boxwood Scale, 12 inches long, one side rounded. the other flat, with the
following scales, the graduations of which are all brought to the edge:
$\frac{1}{16}$, $\frac{1}{8}$, $\frac{3}{16}$, $\frac{1}{4}$, $\frac{3}{8}$, $\frac{1}{2}$, $\frac{5}{8}$, $\frac{3}{4}$, $\frac{7}{8}$, 1, 1¼, 1½, 1¾, 2, 2½ and 3 inches to the
foot, the first division of each scale subdivided into 12 parts, each.....   1 25

615.—Same as No. 614, but has the first division of each scale subdivided into
10 parts, each.................................   ....   1 25

TRIANGULAR BOXWOOD SCALES.

621.

No.   Price.

620.—Triangular Scale of Boxwood, 24 inches long, graduated 10, 20, 30, 40, 50 and 60 to the inch ; or, 20, 30, 40, 50, 60 and 80, to the inch............ $5 00

621.—Triangular Scale of Boxwood, 12 inches long, graduated 10, 20, 30, 40, 50, and 60 to the inch; or, 20, 30, 40, 50, 60 and 80 to the inch............. 2 00

622.— Do.   do. 12 inches long, graduated 100, 200, 300, 400, 500, 600 to the foot each 2 00

623.— Do.   do. 6 inches, graduated same as No. 620. 1 50

624.—Triangular Scales of Boxwood for Off-sets, 2 inches long, 10, 20, 30, 40, 4o. 50 and 60 parts...................... 75

625.—Triangular Scales of Boxwood, 24 inches long, graduated $\frac{3}{32}$, $\frac{3}{16}$, $\frac{1}{8}$, $\frac{1}{4}$, $\frac{3}{8}$, $\frac{1}{2}$, $\frac{3}{4}$, 1, $1\frac{1}{2}$, 3 inches and 16ths to the foot............................... 5 00

626.— Do. do.   do.   12 inches long. 2 00

627.— Do. do.   do.   6 do. 1 50

TRIANGULAR GERMAN SILVER SCALES.

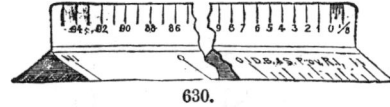

630.

630.—Triangular Scale of German silver, silver plated, 12 inches long, graduated $\frac{1}{8}$, $\frac{1}{4}$, $\frac{3}{8}$, $\frac{1}{2}$, $\frac{3}{4}$, and 1 inch to the foot, each,..................... $6 00

631.—Triangular Scale of German Silver, silver plated, 12 inches long, graduated $\frac{1}{8}$, $\frac{1}{4}$. $\frac{3}{32}$, $\frac{3}{16}$, $\frac{3}{8}$, $\frac{3}{4}$, $\frac{1}{2}$, 1, $1\frac{1}{2}$ and 3 inches to the foot, each ,...... $6 00

632.—Triangular Scale of German Silver, silver-plated, 12 inches long, graduated 10, 20, 30, 40, 50 and 60 to the inch, each,....................... 6 00

633.—Guard for Boxwood Scale, white metal. The use of this Guard, prevents all errors 50

635.—Boxwood Gunter Scales, 12 in. long.................................... 75

636.— do. Gunter Scales, 24 in. long.................................... 1 00

637.—Polar Planimeter, with printed instructions 35 00

By means of Amsler's Polar Planimeter a person entirely ignorant of Geometry may ascertain the area of any planimetrical figure, no matter how irregular its outlines may be, more correctly, and in much shorter time, than the most experienced Mathematician could calculate it.

The management of the instrument can be easily learned in half an hour, and in size it is no larger than a two-foot folding rule.

The Planimeter indicates square feet or square inches, and acres for surveying.

PAPER SCALES.

640.—Paper Scale, printed on card-paper, $1\frac{1}{4}$ inch wide, 12 inches long, graduations on one edge inches and 10ths, and the other feet and 100ths .... $ 10

641.—Paper Scale, same as 640, one edge 20 parts to the inch, the other edge 40, ... 10

642.—Paper Scale, same as 641, one edge inches and sixteenths, the other edge inches and forty-eighths,.. 10

643.—Paper Scales, printed on card-paper, 19 inches long, for architects and engineers, in sets of 6 scales...................... .... 1 00

No.   Price.

644.—Series A contains 6 scales, one each divided to ¼, ½, ¾, 1, 1½, and 3 inches to the foot, per set $1 00

645.—Series B contains 6 scales, one each, divided to $\frac{3}{32}$, ⅛, $\frac{3}{16}$, $\frac{5}{16}$, ⅜, and ⅞ inch to the foot, per set 1 00

646.—Series C contains 6 scales, one each, divided to 10, 20, 30, 40, 50 and 60 parts to the inch, per set 1 00

Single Scale of any of the above series, A, B, C—each scale,......... ...   20

647.—Paper Scales, same as 643, divided either to ⅝, 1⅛, 1¼, 1⅜, inches to the foot, each................... 20

The advantages of these scales are—they expand and contract nearly the same as drawing-paper, do not soil the work, and distances can be set off from them without the use of dividers.

We Manufacture, to order, scales to any Divisions, in Ivory Boxwood or Rubber.

STEEL RULES, GAUGES, SQUARES, CALIPERS, FOR MACHINISTS, STRAIGHT EDGES, &c.

STANDARD STEEL RULES.

Always give graduation when ordering these goods.

				Price
650.—Steel Rule 1 inch long......................................				$ 25
651.—	Do.	2	do.	40
652.—	Do.	3	do.	50
653.—	Do.	4	do.	75
654.—	Do.	6	do.	1 00
655.—	Do.	9	do.	1 50
656.—	Do.	12	do.	2 00
657.—	Do.	18	do.	3 00
658 —	Do.	24	do.	4 00
659.—	Do.	36	do.	8 00
660.—	Do.	48	do.	12 00

The rules in this list are divided five ways in parts of inches as follows:

No. 1 Graduations.	No. 2 Graduations.	No. 3 Graduations.
1st cor. 10, 20, 50, 100	10, 20, 50, 100	16, 32, 64
2d cor. 12, 24, 48	12, 24, 48	16
3d cor. 16, 32, 64	16, 32, 64	16
4th cor. 14, 28	8	8

No. 4 Graduations.	No. 6 Graduations.	
1st cor. 64	1st cor. 32 whole length.	
2d cor. 32	2d cor. 48	do.
3d cor. 16	3d cor. 50	do.
4th cor. 8	4th cor. 64	do.

STEEL RULES NO. 5 GRADUATION.

No.   Price.

661.—12 in. steel Rule of No. 5 graduation............ .... $3 00

662.—24 in. Do. do. do. do. do, ... 6 00

No. 5 Graduations.

1st cor. 16, 32, 64

2d cor. 11, 14, 15, 17, 18, 19, 20, 21, 22, 23, 24, 25

3d cor. 26, 27, 28, 29, 30, 31, 33, 34, 35, 36, 37, 38

4th cor. 39, 40, 41, 42, 43, 44, 45, 46, 47, 48, 49, 50, 100

FRENCH STANDARD STEEL RULES.

664.—Steel Rule, 1 Metre long...$10 00

665.— Do. $\frac{1}{2}$ do. 4 00

666.— Do. $\frac{3}{10}$ do. 2 50

667.— Do. $\frac{1}{5}$ do. 1 75

668.— Do. $\frac{1}{10}$ do. 0 85

669,— Do. $\frac{1}{20}$ do. 0 45

They are divided on three edges to millimetres, and on one edge to fifths of millimetres.

Steel Rule, $\frac{1}{5}$ Metre, send by mail on receipt of......... .... 2 00

Do. $\frac{1}{10}$ do. do. do. do. 0 90

Do. $\frac{1}{20}$ do. do. do. do. 0 50

TRIANGULAR STEEL RULES.

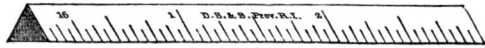

670.

670.—Triangular Steel Rule, 3 inches long..................................... $ 60

671.— Do. do. do, 4 do. do. 80

672.— Do. do. do. 6 do. do. 1 20

673.— Do. do. do. 12 do. do. 3 00

Graduations.

16, 64, 100 to the inch whole length.

16, 32, 64 Do. do. do.

20, 50, 100,—12, 24, 48,—16, 32, 64 to the inch.

The 12 in. are divided only as follows: 8, 10, 12, 14, 16, 20, 24, 28, 48, 50, 64, 100 to the inch.

SQUARE STEEL RULES.

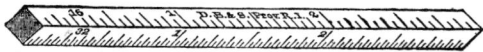

675.

675.—Square Steel Rule, 3 inches long .... $ 50

No.	Price.
676.—Square Steel Rule, 4 inches long	$ 75
677.— Do. do. 6 do. do.	1 00

Graduations.

8, 16, 32, 64 to the inch whole length.

16, 32, 64, 100   Do.   do.   do.

16, 64, 50, 100   Do.   do.   do.

STANDARD STEEL STRAIGHT EDGES.

Of same width and thickness as Standard Rules.

680.— 6 in. long, 1 in. wide, $\frac{1}{10}$ in. thick	$ 75
681.— 9 in. Do, 1⅛ in. do. $\frac{1}{10}$ in. do.	1 12
682.—12 in. Do. 1¼ in. do. $\frac{1}{10}$ in. do.	1 50
683.—18 in. Do. 1½ in. do. ⅛ in. do.	2 25
684.—24 in. Do. 2 in. do. ⅛ in. do.	3 00
685.—36 in. Do. 2⅜ in. do. ⅛ in. do.	6 00
686.—48 in. Do. 3 in. do. ⅛ in. do.	9 00

STEEL STRAIGHT EDGES.

FOR DRAUGHTSMEN.

690.—15 in. long, 1¼ in. wide, $\frac{1}{20}$ in. thick	$1 14
691.—18 in. Do. 1½ in. do. $\frac{1}{20}$ in. do.	1 62
692.—24 in. Do. 1½ in. do. $\frac{1}{18}$ in. do.	2 16
693.—30 in. Do. 1¾ in. do. $\frac{1}{18}$ in. do.	3 15
694.—36 in. Do. 2 in. do. $\frac{1}{16}$ in. do.	4 32
695.—42 in. Do. 2¼ in. do. $\frac{1}{16}$ in. do.	5 67
696.—48 in. Do. 2½ in. do. $\frac{1}{14}$ in. do.	7 20
697.—60 in. Do. 2¾ in. do. $\frac{1}{12}$ in. do.	9 90

SHRINK RULES FOR PATTERN MAKERS.

700.—24¼ inch Steel Rule, shrink on one side and standard on the other, divided on each side, 10, 20, 50, 100, 12, 24, 48, 16, 32, 64, parts to inch	$4 50
701.—24¼ inch Steel Rule, shrink on both sides, No. 1 graduation	4 50
702.—24¼ inch Boxwood Rule, shrink on both sides, No. 1 graduation	3 00

STANDARD SCALES AND CENTRE GAUGES—STEEL.

705.—36 inch Steel or Standard Yard, full, divided	$ 3 00
706.—French Standard Metre, divided on 3 edges to millimetres, and one edge to 5ths of millimetres	10 00
707.—Centre Gauge for lathes, also for screw tools	0 50

GRADUATED STEEL SQUARE FOR MACHINISTS.
NOT HARDENED.

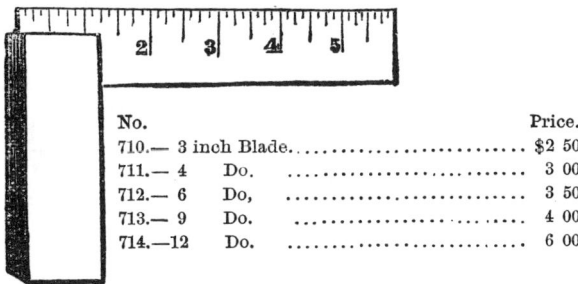

No.		Price.
710.— 3 inch Blade	....	$2 50
711.— 4	Do.	3 00
712.— 6	Do, ...	3 50
713.— 9	Do.	4 00
714.—12	Do.	6 00

PATENT HARDENED CAST STEEL TRY SQUARES.
FOR MACHINISTS.

No.	Price.	No.	Price.
715.—1½ inch	$2 50	721.—15 inch	$15 00
716.—3 "	3 50	722.—18 "	20 00
717.—4½ "	4 50	723.—24 "	30 00
718.—6 "	6 00	724.—30 "	40 00
719.—9 "	9 00	725.—36 "	50 00
720.—12 "	12 00		

THIN STEEL SQUARES.
FOR MACHINISTS AND DRAUGHTSMEN.

726.—2 in. blade, 1-20 in. thick,	$1 50	729.—6 in. blade, 1-14 in. thick,	$3 50
727.—3 " 1-16 "	2 00	730.—8 " 1-12 "	4 50
728.—4 " 1-14 "	2 50	731.—10 " 1-10 "	5 50

The 2 and 3-inch are divided to 16ths and 64ths on one side, and 32ds and 64ths on the other.

The 4, 6, 8 and 10-inch Squares are divided on both sides to 16ths and 32ds of inches.

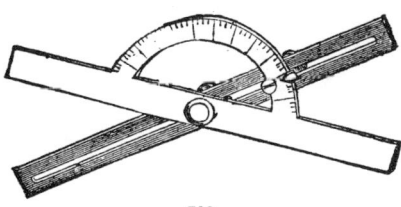

732.

BEVEL PROTRACTORS.
With sliding arm, and half circle divided to degrees.

732.— 6 in. sliding arm	....	$6 50
733.—10 in. Do.	....	7 00

STEEL CALIPERS.

No.   Price.

735.—Hardened cast steel pocket Vernier Caliper, Vernier to 1000ths of inches. one edge to Millimetres. $5 00

736.— Do.   do.   do.   in Morocco Case 6 00

737.— Do.   do.   do.   with Adjusting Screw 6 00

738.— Do·   do.   do.   with Adjusting Screw in Morocco Case...... 7 00

739.—Improved Vernier Caliper, vernier reads to 1000ths of inches inside and outside Calipers, and points to transfer the distance, with dividers ; in morocco case ; 6-inch, ... 25 00

740.—   Do.   12-inch, ... 30 00

CALIPER SQUARES.

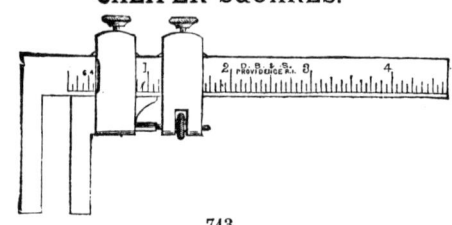

743.

741.—2 in. without adjusting screw,... $3 50

742.—4 in.   "   "   " 4 50

743.—2 in. with adjusting screw like cut,............................... .... 4 25

744.—4 in.   "   "   "   " 5 50

UNIVERSAL, OR CENTRE SQUARES.

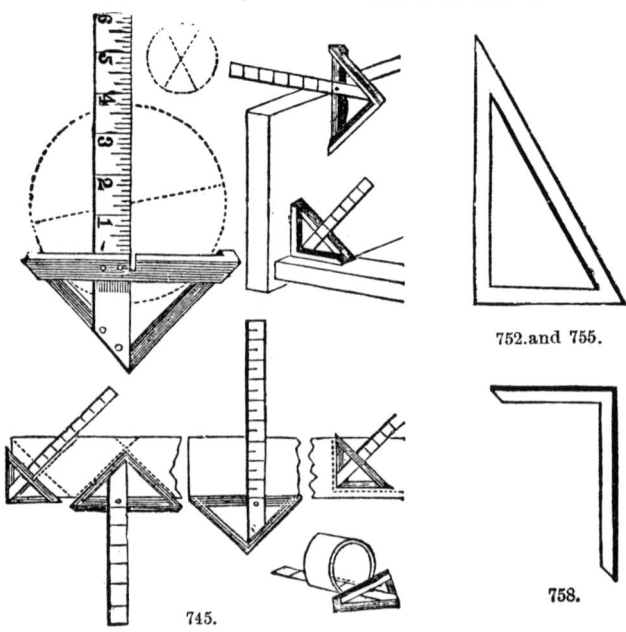

752.and 755.

745.

758.

No.							Price.
745.— 4 in.,							$3 00
746.— 6 "							3 00
747.— 8 "							4 00
748.—10 "							5 00
749.—12 "							6 00

OPEN STEEL TRIANGLES.

750.—Angles, 45,45 and 90 degrees. Large size, sides 8, 8 and 11 1-4 in.,3-4in. wide $4 50

751,—Small size, sides,5, 5 and 7 1-16 in., 5-8 in. wide, ... 3 50

752.—Angles, 30, 60 and 90 degrees. Large size, sides 6, 10 3-8 and 12 in., 3-4 in. wide, ... 4 50

753.—Small size, sides 3 1-2, 6 1-16 and 7 in., 5-8 in wide, Thickness 1-16 inch.. 3 50

GERMAN SILVER TRIANGLES AND SQUARES.

No.							Price.
754.—German Silver Triangle—30, 60 and 90 deg. perpendicular, 6 to 7 inch.							$2 50
755.— do	do	do	30, 60 and 90 deg.	do	9 to 10 inch.		4 00
756.— do	do	do	45, 45 and 90 deg.	do	4 to 5 inch.		2 25
757.— do	do	do	45, 45 and 90 deg.	do	6 to 7 inch.		3 50
758.— do	do	Squares, perpendicular, 6 to 7 inches				....	75

Other sizes of Steel and German Silver Triangles made to order.

STRAIGHT EDGES, TRIANGLES, T SQUARES, &c.

(For steel straight edges, see numbers, 680 to 697.)

Straight Edges.

RUBBER AND WOOD.

No.							Price.
760.—Hard Rubber Straight Edges, 42 inches long, ...							$2 75
761.— Do.	do.	36	do.			....	2 20
762.— Do.	do.	30	do.			....	1 80
763.— Do.	do.	24	do.			....	1 25
764.— Do.	do.	18	do.			....	75
765.—Whitewood Straight Edges, 20 inches long, with one edge bevelled,						.....	25
766.— Do.	do.	30	do.	do.	do.	....	35
767.— Do.	do.	40	do.	do.	do.	....	55
768.— Do.	do.	50	do.	do.	do.	....	70
769.— Do.	do.	60	do.	do.	do.	....	1 00
770.—Polished Rosewood or Satinwood Straight Edges, 20 inches long,						....	50
771.— Do.	do.		do.	30	do.	....	75
772.— Do.	do.		do.	40	do.	....	1 00
773.— Do.	do.		do.	50	do.	....	1 50
774.— Do.	do.		do.	60	do.	....	2 50
775.— Do.	do.		do.	72	do.	....	4 00

TRIANGLES.

Rubber, Rosewood, Satinwood or Whitewood.

Hard Rubber Triangles, 30 × 60°.

785.   798.

No.		Price.
780.—3 inch .... $ 25	787.—10 inch .... $ 65	
781.—4 " 25	788.—11 " 75	
782.—5 " 30	789.—12 " 90	
783.—6 " 35	7.0.—13 " 1 00	
784.—7 " 40	791.—14 " 1 25	
785.—8 " 50	792.—15 " 1 50	
786.—9 " 60	793.—16 " 1 75	

Hard Rubber Triangles 45°.

794 —3 inch .... $ 35	801.—10 inch $1 00	
795.—4 " 35	802.—11 " 1 25	
796.—5 " 40	803.—12 " 1 50	
797.—6 " 45	804.—13 " 1 75	
798.—7 " 55	805.—14 " 2 00	
799.—8 " 70	806.—15 " 2 25	
800.—9 " 80	807.—16 " 2 50	

Cross Section Triangles.

Hard Rubber.

808.—Cross Section Triangles, set of seven Cross Section Triangles made of hard rubber as follows, ¼ to 1, ½ to 1, ¾ to 1, 1 to 1, 1¼ to 1, 1½ to 1, 2 to 1, per set, ... $4 25

Single Triangles of set No. 808, each........... 75

Batter Slopes.

809.—Set of three forms of hard rubber for Batters of walls and rock, giving the following slopes, 1 in. 4, 1 in. 5, 1 in. 6, 1 in. 8, 1 in. 10, 1 in. 12, per set, ... $2 00

Single forms of set No. 809, containing any two slopes, each............. 75

Rosewood, Satinwood, and Whitewood Triangles.

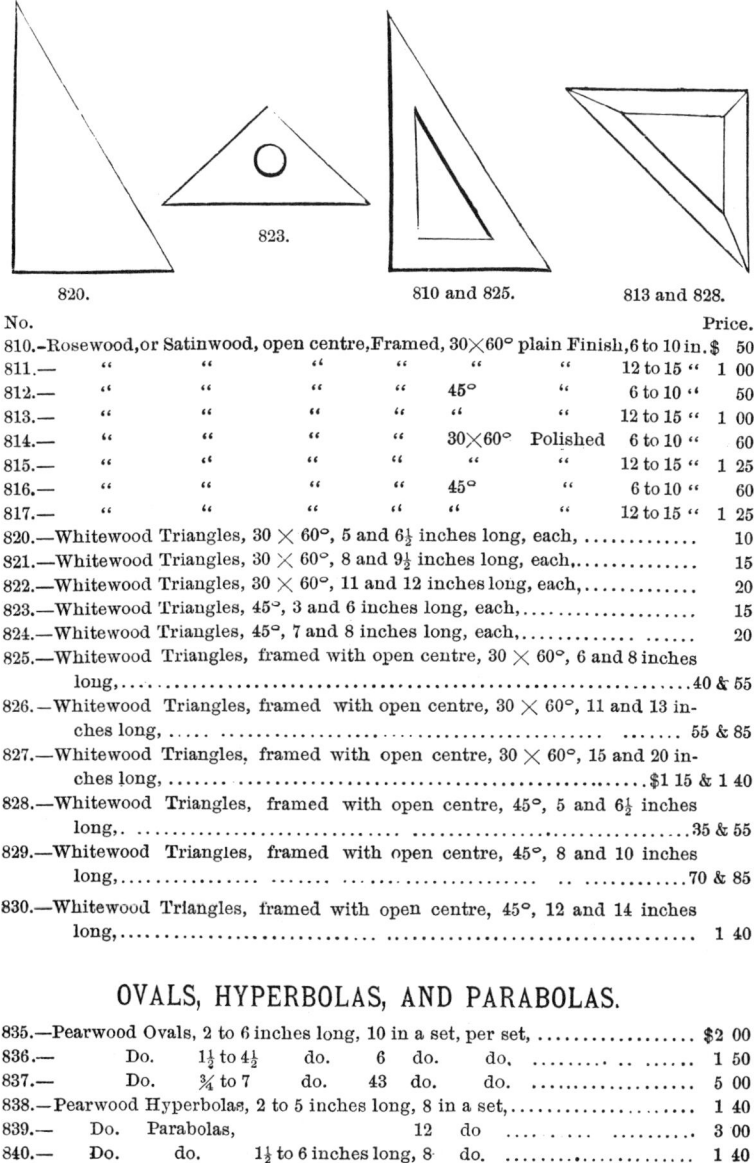

823.

820.   810 and 825.   813 and 828.

No.   Price.

810.-Rosewood, or Satinwood, open centre, Framed, 30×60° plain Finish, 6 to 10 in. $ 50
811.—   "   "   "   "   "   "   12 to 15 "   1 00
812.—   "   "   "   "   45°   "   6 to 10 "   50
813.—   "   "   "   "   "   "   12 to 15 "   1 00
814.—   "   "   "   "   30×60°   Polished   6 to 10 "   60
815.—   "   "   "   "   "   "   12 to 15 "   1 25
816.—   "   "   "   "   45°   "   6 to 10 "   60
817.—   "   "   "   "   "   "   12 to 15 "   1 25
820.—Whitewood Triangles, 30 × 60°, 5 and 6½ inches long, each, ... 10
821.—Whitewood Triangles, 30 × 60°, 8 and 9½ inches long, each, ... 15
822.—Whitewood Triangles, 30 × 60°, 11 and 12 inches long, each, ... 20
823.—Whitewood Triangles, 45°, 3 and 6 inches long, each, ... 15
824.—Whitewood Triangles, 45°, 7 and 8 inches long, each, ... 20
825.—Whitewood Triangles, framed with open centre, 30 × 60°, 6 and 8 inches
        long, ...40 & 55
826.—Whitewood Triangles, framed with open centre, 30 × 60°, 11 and 13 in-
        ches long, ... 55 & 85
827.—Whitewood Triangles, framed with open centre, 30 × 60°, 15 and 20 in-
        ches long, ...$1 15 & 1 40
828.—Whitewood Triangles, framed with open centre, 45°, 5 and 6½ inches
        long, ...35 & 55
829.—Whitewood Triangles, framed with open centre, 45°, 8 and 10 inches
        long, ...70 & 85
830.—Whitewood Triangles, framed with open centre, 45°, 12 and 14 inches
        long, ... 1 40

OVALS, HYPERBOLAS, AND PARABOLAS.

835.—Pearwood Ovals, 2 to 6 inches long, 10 in a set, per set, ... $2 00
836.—   Do.   1½ to 4½   do.   6   do.   do, ... 1 50
837.—   Do.   ¾ to 7   do.   43   do.   do. 5 00
838.—Pearwood Hyperbolas, 2 to 5 inches long, 8 in a set, ... 1 40
839.—   Do.   Parabolas,   12   do 3 00
840.—   Do.   do.   1½ to 6 inches long, 8 do. 1 40

T SQUARES.

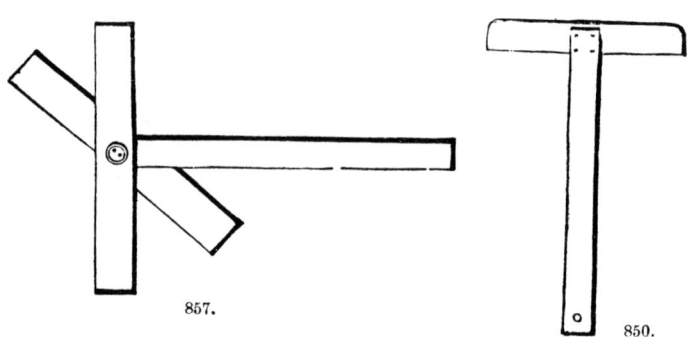

857.

850.

No.						Price.
850.—Pearwood T Square, fixed head, blade 15 inches long, each, ...						$ 35
851.—	Do.	do.	do. 20	do.	do.	45
852.—	Do.	do.	do. 25	do.	do.	55
853.—	Do.	do.	do. 30	do.	do.	75
854.—	Do.	do.	do. 35	do.	do.	85
855.—	Do.	do,	do. 40	do.	do.	95
856.—	Do.	do.	do. 50	do.	do.	1 40
857.—	Do.	double	do. 20	do.	do.	1 15
858.—	Do.	do.	do. 25	do.	do.	1 35
859.—	Do.	do.	do. 30	do.	do.	1 50
860.—	Do.	do.	do. 35	do.	do.	1 60
861.—	Do.	do.	do. 40	do.	do.	1 70
862.—	Do.	do.	do. 50	do.	do.	2 00
865.—Rosewood or Satinwood T Square, Fig. 850, polished, 30-inch, ...						1 75
866.—	Do.	do.	do.	do.	40 do.	2 50
868.—Rosewood or Satinwood T Square, polished Fig. 857, 30-inch, ...						2 75
869.—	Do.	do.	do.		do. 40 do.	3 50

NICKEL PLATED LIMB PROTRACTORS.

(Superior.)

870.—20 to 24 inches			$18 00
871.—30 inches.			20 00
872.—36	"	. . . .	22 00
873.—48	"	. . . .	25 00

Divided to half degrees with a vernier reading to five Minutes.

Nos. 865 to 873 inclusive are our own Manufacture.

IRREGULAR CURVES.

WOOD.

No.   Price.

900.—Whitewood Irregular Curves, 5 to 15 inches long, various patterns, each, $ 20

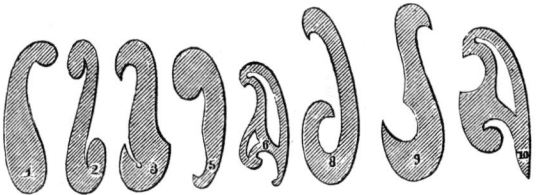

901.—Whitewood Irregular Curves, of superior quality 6½ to 10 inches long,
various patterns, each,.. 25

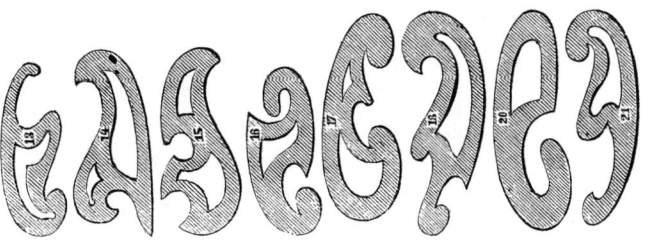

902.—Whitewood Irregular Curves, of superior quality, 8 to 12 inches long,
various patterns, each, ... 35

No.   Price.

903.—Whitewood Irregular Curves, of superior quality, 13 to 17 inches long, various patterns, each,.. 50

RAILROAD REGULAR CURVES.

CARD BOARD WOOD AND RUBBER.

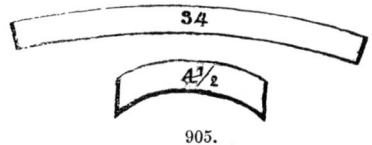

905.

No.   Price.

905.—Railroad Curves, of card board. Set of 24 curves, from $1\frac{1}{2}$ to 24 inches radii, in wood box, per set, ... $4 50

906.—Railroad Curves, of card board. A set of 50 curves, from $1\frac{1}{2}$ to 120 inches radii, in wood box, per set, ... 9 00

907.—Railroad Curves, of card board, A set of 100 curves from $1\frac{1}{2}$ to 240 inches radii, in wood box, per set,... 15 00

908.—Railroad Curves, of card board. A set of 25 curves, from 30 minutes to 7 degrees by every 15 minutes, cut to a scale of 400 feet to the inch, in wood box, per set,... 5 00

909.—Railroad Curves of card board. A set of 70 curves, from 25 minutes to 4 degrees by every 5 minutes, and from 4 degrees to 10 degrees by every 15 minutes, in wood box, per set,.................................... 15 00

910.—Railroad Curves, of wood. Set of 43 curves, of radii from $3\frac{1}{2}$ to 200 inches, pet set,... 9 00

911.—Railroad Curves, of hard rubber. Set of 10 curves, 12 to 120 inches radii, in case, per set,... 7 75

912.—   Do.   do.   17 curves, 12 to 60 inches radii, pet set,.............. 13 25

913.—Railroad Curve Protractor, of horn, 8 inches diameter, having laid off on it 33 curves, from $\frac{1}{2}$ degree to 8 degrees, with a radii of 400 feet to the inch, each,... 2 00

HARD RUBBER CURVES.

925.

No.
925.—Hard Rubber Curves:

No.					Price.	No.					Price.
No.	1.	5½ inches long	.....	each	35	No.	19.	8 inches long	.....	each	50
"	2.	5½	"	..... "	35	"	20.	10½	"	..... "	50
"	3.	9	"	..... "	50	"	21.	7½	"	..... "	45
"	4.	9	"	..... "	5⁰	"	22.	5	"	..... "	25
"	5.	6	"	..... "	40	"	23.	6	"	..... "	40
"	13.	9	"	..... "	50	"	24.	9	"	..... "	60
"	14.	7½	"	..... "	35	"	25.	7	"	..... "	40
"	15.	8½	"	..... "	45	"	26.	5½	"	..... "	35
"	16.	4¾	"	..... "	35	"	27.	12	"	..... "	1 00
"	17.	9	"	..... "	35	"	28.	12	"	..... "	3 00
"	18.	8	"	..... "	40						

PARALLEL RULES, &c.

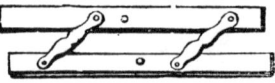 

No.	950.		957, 960, 965 and 968.	Price

950.—Parallel Rulers, ebony, brass mounted, 6 inches long, each,............ . $0 25

951.—	Do.	do.	do.	9	do.	do.	50
952.—	Do.	do.	do.	12	do.	do.	75
953.—	Do.	do.	do.	15	do.	do.	1 00
954.—	Do.	do.	do.	18	do.	do.	1 25
955.—	Do.	do.	do.	24	do.	do.	2 00
956.—	Do.	German Silver mounted, 12			do,	do.	1 25

957.—Parallel Rulers, all German Silver, on Rollers, 12 inches,............... 10 00

958.—	Do.	do.		do.	15	do.	13 00
959.—	Do.	do.		do.	18	do.	15 00
960.—	Do.	all Brass		do.	9	do.	5 00
961.—	Do.	do.		do.	12	do.	6 50
962.—	Do.	do.		do.	15	do.	8 50
963.—	Do.	do.		do.	18	do.	10 00
964.—	Do.	do. Nickel Plated	do.		9	do.	6 50
965.—	Do.	do. do.	do. do.		12	do.	8 00
966.—	Do.	do. do.	do. do.		15	do.	10 00
967.—	Do.	do. do.	do. do.		18	do.	12 00
968.—	Do.	Ebony,	on Rollers		9	do.	2 75
969.—	Do.	do.	do.		12	do.	3 25
970.—	Do.	do.	do.		15	do.	4 00
971.—	Do.	do.	do.		18	do.	5 00
972.—	Do.	do.	do. Graduated Edges 12 inches.				5 00
973.—	Do.	do.	do.	do.	15	do.	6 00
974.—	Do.	do.	do.	do.	18	do.	7 50

Rules 957 to 971 inclusive are our own Manufacture, and are of superior quality.

SECTION LINERS.

975.—Bergner's Patent Section Liner, in Morocco Case........................ 12 00

976.—Queen's Section Liner, in Paper Box.................................... 2 00

977.—Marion's Section Liner, in Paper Box, German Silver Slide and Screws..
    with polished satinwood Triangle, and Ruler............. .... 1 50

977.

CENTROLINEAD.

No.   Price.

978.—Centrolinead, of wood, for perspective drawing, arms 24 inches, per pair, $5 00
979.—   Do.   do.   do.   do. 36   do.   ..   6 50

PANTOGRAPHS.

980.—Pantograph of common wood arms, ... $3 00
981.—   Do.   pearwood, arms 22 inches long,......................... 5 50

FASTENING TACKS AND HORN CENTRES.

| 1001. | 1006. | 1012. | 1013. | 1014. |

1001.—Fastening Tacks, of Brass, heads round, ¼ inch diameter,......per doz. $ 25
1002.—   Do.   do.   do. $\frac{7}{16}$ do.   ......   do.   35
1003.—   Do.   of German Silver, heads rounded, ¼ in. diam.,   do.   50
1005.—   Do.   do.   do. $\frac{7}{16}$ do. ......do.   60
1066.—   Do.   do. flat heads, $\frac{5}{16}$ in. diam. very superior,   do.   70
1007.—   Do.   do.   do. $\frac{7}{16}$ do.   do. ......   do.   80
1608.—   Do.   do.   do.   ½   do.   do. ......   do.   1 00
1009.—   Do.   do.   do.   ⅝   do.   do. ......   do.   1 25
1010.—   Do.   of Steel, round heads, $\frac{7}{16}$ in.diam. per doz.   55
1011.—   Do.   Steel Tacks, Swiss $\frac{7}{16}$ "   do.   75
1012.—   Do.   of Brass, right angled.............. do.   75
1013.—Horn Centre, each................................ do.   15
1014.—   Do.   with German Silver Rim, each....... do.   35
1015.—German Silver, centre, with Handle ....each   30

BOXWOOD COMBINATION RULE.

1020.

No.   Price.

1020.—Combination Rule, One Foot, Two Fold, boxwood. This is the most convenient and useful pocket-rule ever made ; it combines in itself a car-
penter's Rule, Spirit Level, Square Plumb, Bevel, Indicator Brace,
Scale, Draughting Scale of equal parts, T Square, Protractor, Right
angle Triangle, and with a straight edge can be used as a Parallel
Ruler, all the parts of which, in their separate application, are perfectly
reliable,.. $3 00

POCKET RULES.

Boxwood, One Foot, Pocket.

1021.—Round joint, ⅝ in. wide..$ 25
1022.—Arch joints, Edge plates, ⅝ in. wide.................................... 50
1023.—Square "   Bound, 8th, 10th, 12th and 16th 1 00

Boxwood, Two Foot, Six Fold.

1025.—Boxwood, 2 foot, 6 fold, graduated 8th, 10th,1 00th and 16th inches, for
engineers... 1 25

Boxwood, Two Foot, Narrow, Four Fold.

1026.—Round joints, 1 in. wide.. 30
1027.—Arch   "   "   50
1028.—Square   "   Bound, 1 inch wide, with drafting scales................ 1 25

Boxwood, Two Foot, Broad, Four Fold.

1029.—Square joints, Edge plates 75
1030.—Arch   "   "   8th, 10th, 12th and 16th...................... 90
1031.—Double Arch joints, 8th, 10th, 12th and 16th 1 00
1032.—Square joints, Bound, 8th, 10th and 16th, and Drafting Scales.......... 1 50

Boxwood, Two Foot, Two Fold.

1033.—Arch joint, Gunters slide, Engineers, 8th, 10th, 16th and 100th, Drafting
and Octagonal Scales 1 50
1034.- Square joint, Plain Slide, 8th, 10th and 16th, Octagonal Scales.......... 75

Boxwood Caliper Rules.

1035.—2 fold, 6 in., plain, ⅞ in. wide, 8th, 10th and 16th...................... 60
1036.—4 " 12 " Bound, 1   "   "   " 1 67

IVORY, ONE FOOT, POCKET.

1040.—Round joint, Brass, ½ in. wide 75
1041.—   "   German Silver, ½ in. wide.............................. 1 00
1042.—Arch joints, Edge plates, German Silver, ⅝ in. wide, graduations in 8th,
10th, 12th, 16th, and 100th, for Engineers 1 75
1043.—Square joint, Bound, German Silver, ⅝ in. wide, graduation in 8th, 10th,
12th, 16th, and 100th, for Engineers........ .... 2 35
1044.—Arch joint, Bound, German Silver, 8th, 10th, 12th, 16th, and 100th on
edge, ⅝ in. wide 2 50

IVORY, TWO FOOT, FOUR FOLD.

No.   Price.

1045.—Arch joint, Edge plates, 1 inch wide, German Silver, graduated to 8th, 10th, 16th and 100th, and Drafting Scales............................. $5 50

1046.—Square joint, Edge plates, ¾ in. wide German Silver, graduated to 8th, 10th, 18th and 100th, and Drafting Scale............................. 4 50

IVORY CALIPER RULES.

1047.—2 fold, 6 in., German Silver.. $1 50
1048.—2 " " Bound ".. 2 50
1049.—4 " 12 in., Plain " ⅞ in. wide, 8th, 10th, 12th and 16th...... 3 18
1050.—4 " " Bound " " " " 4 00

BENCH RULES, PATTERN MAKERS' SHRINK RULES, YARD-STICKS, GAUGING AND WANTAGE RODS, MEASURING RODS, BOARD MEASURES, &c.

1051.—Bench Rule Boxwood, Bound, 8th and 16th, 24 in. long................. $1 50
1052.—Pattern Makers Shrinkage Rule, Boxwood, 8th and 16th in., 24½ in. long 1 25
1053.— " " two fold, Boxwood, Triple plated, Edge plates, 8th and 16th, 24½ in. long. 1 50
1054.—Yard Stick, Capped Ends.. 30
1055.—Gauge Rods, 3 feet, 120 gals., each.................................... 70
1056.— " 4 " 300 " " 75
1057.—Wantage Rods, Boxwood, 8 scales...................................... 75
1058.— " " 12 " 1 00

Measuring Rods of Sprucewood.

1060.—6 feet long, ¾ in. square, ends tipped, Divided to feet, inches and ¼ in. 1 00
1061.—8 " ⅞ " " " 10th and 100th... 1 10
1062.—8 " ⅞ " " " inches and ⅛ in.. 1 10
1063.—10 " ⅞ " " " 10th and 100th... 1 25
1064.—10 " ⅞ " " " inches and ⅛ in.. 1 25
1065.—13 " 1¾ by ⅞ in., feet, 10th and 20th, painted.................... 5 00

Board Measures.

1066.—Board Stick, Octagon (8 to 23 feet), Brass Cap, 3 feet long.............. 1 00
1067.—Walking Cane Board Measure, Octagon, Hickory, Cast Brass Head and Tip, 8 lines (9 to 16 feet), 3 feet long................................. 1 00

SPIRIT LEVELS, POCKET COMPASSES, PLUMBOBS AND MAGNETS.

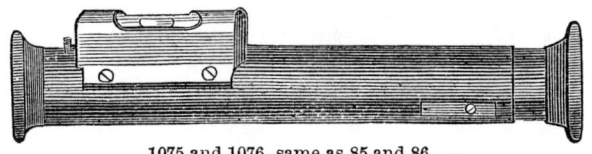

1075 and 1076, same as 85 and 86.

No.   Price.
1075.—Locke's Hand Level, made of German Silver (*See* Cut, p. 187.) $12 00
1076.—   "   "   " Brass   ("   ) 10 00

1079.

1079.—Pocket Levels, mounted in Brass, 3 inches long....................... $ 75
1080.—   "   "   " 6   " 1 50
1081.—   "   "   " 9   "   ....: 2 ٢5
1082.—   "   "   " 12   " 3 00
1085.—Spirit Level, Arch Top plate, two side viens, polished................. 1 00
1086.—Spirit Level, Plumb and Level, Arch top plate, two side veiws, polished,
      24 to 30 inches.. 1 50

1087.

1087.--Spirit Level, Patent adjustable Plumb and Level, Ground Vials, Arch
      top plate, two side views, polished and tipped, 26 to 30 inch........... 2 50

1088.

1088.—Spirit Level, Patent Adjustable Plumb and Level, Ground Vials, Arch
      top plate, two ornamental brass tipped side views, polished and tipped,
      26 to 30 inch.. 4 00

1090.

1090.—Mason's Plumb and Level, Patent Adjustable (Ground Vial), 3¾ in. wide,
      42 in. long... 3 00

Section of Patent Adjustable Plumbs and Levels.

Adjustment of the Patent Improved Adjustable Plumbs and Levels. The
bubble tube in the Level, or Plumb, is attached to the brass top-plate
above it, at one end by hinge, and at the opposite end by an adjusting
screw, which passes down through a flange on the metallic case. Between this flange and the top-plate, is inserted a stiff spiral spring, and
by driving or slacking the adjusting screw, should occasion require it,
the bubble tube can be adjusted to a position parallel with the base of
the Level.

The simplicity of this method of adjustment, will commend itself to every
mechanic, and is reliable under all circumstances.

No.			Price.
1095.—Level Bulbs, unmounted, 2 to 4 inches..................................			$ 25
1096.—	"	" 4 to 6 "	35
1097.—	"	" ground vials, 2 to 6 inches, 50c. to	2 50

Horse Shoe Magnets.

1100.—2 inch..		15
1101.—2½ inch...		.20
1102.—3 "		40
1103.—4 "		50
1104.—5 "		70
1105.—6 "		1 00
1106.—7 "		1 37
1107.—8 "		1 75

IMPROVED TRAMMEL POINTS.

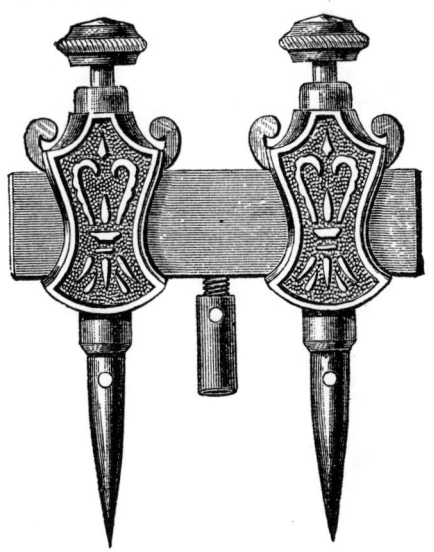

1111.

These tools are used by all who have occasion to strike arcs, or circles, larger than can be done by compass Dividers. They may be used on a straight wooden bar of any length, and when secured in position by the thumb screws, all circular work can be readily laid out. They are made of bronze, and have steel points, either of which can be renewed, and replaced by pencil socket, which accompanies each pair.

1110.—Small (No. 1)...per pair $1 50		
1111.—Medium (No. 2)... "		2 00
1112.—Large (No 3)... "		2 75

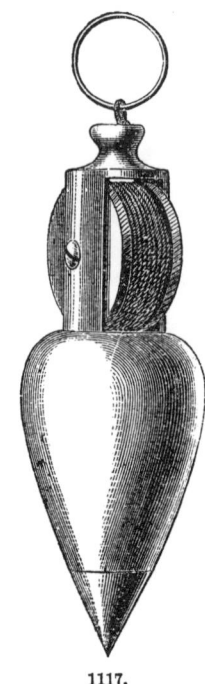

1117.

PATENT ADJUSTABLE PLUM BOBS.

These are constructed with a reel at the upper end, upon which the line may be kept, and by dropping the bob with a slight jerk, while the ring is held in the hand, any length of line may be reeled off. A spring which has a bearing on the reel, will check and hold the bob firmly at any point on the line.

No.		Price.
1114.—Small (No. 1), 8 oz.		$1 75
1115.—Large (No. 2), 13 oz		2 25

BRASS PLUMB BOBS.

W. & L. E. GURLEY, MANUFACTURERS.

No.				Price.
1120.—Steel Point, Screw Heads, 3 oz				$1 00
1121.—	"	"	6 "	1 50
1122.—	"	"	10 "	1 75
1123.—	"	"	14 "	2 25
1124.—	"	"	20 "	2 75
1125.—	"	"	24 "	3 00
1126.—	"	"	32 "	3 50

POCKET COMPASSES.

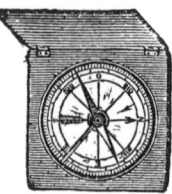

1130.   1139.   1136.

No.				Price.
1130.—Mahogany Case, stop to needle, 1¼ inches Square				$1 50
1131.—	"	"	2 "	2 00
1132.—	"	"	2½ "	2 25
1133.—	"	"	3 "	2 75
1134.—Brass, round, with cover 1½ inches diameter, stop to needle				1 25
1135.—	"	"	" and agate centre to needle		1 75
1136.—Brass, round, watch pattern, stop, agate centre, 1½ inch				1 50
1137.—	"	"	" " 2 "	2 00

No.   Price.

1138.—Brass, round, watch pattern, stop, agate centre, 1½ iuch, with hinged
          cover.. $2 50
1139.—   "   Gilt Charm Compasses to hang to watch guard...........   25
1140.—Pocket Compass, gilt watch pattern, with stop, enameled dial, morocco
          case. London, Superior, 1⅛ inches................................   6 00
1141.—   "   "   1¾ " 8 00
1150.—Prismatic Azimuth Compass, brass, 2¾ in. diam. 18 00
1151.—   "   "   "   "   4   " 22 00

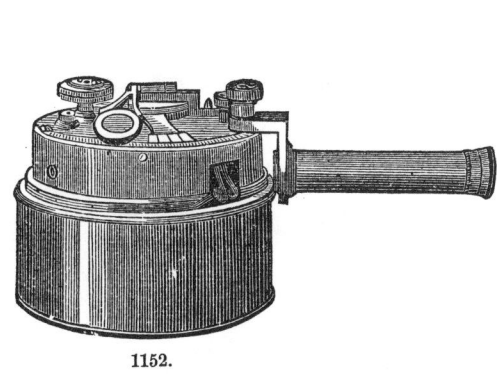

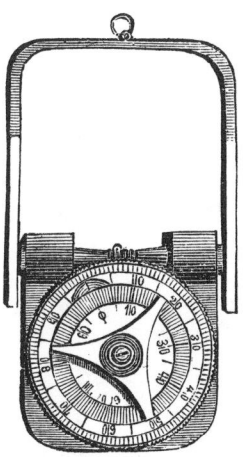

1152.

1154.

1152.—Pocket Sextants, with Telescope, very accurate...................... .. 50 00
1153.—Surveyor's Cross—for right angles........ .... 3 00
1154.—Odometer, for measuring distances by recording revolutions of carriage
          wheel.. 20 00
1155.—Pedometer, for measuring distances walked, watch form and size, Ger-
          man Silver case...................................... .... 17 00
1156.—   "   "   "   Silver case 21 00

MICROSCOPES, &c.

SIMPLE MICROSCOPES, TO FOLD IN CASES.

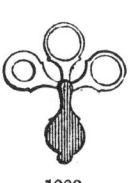

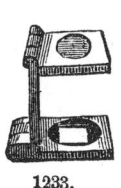

1200 to   1224.   1226.   1229.   1233.
1223.   1237

No.						Price.
1200—	Hard Rubber case & frame,	rounded Rim,	1 double convex lens	¾	in.diam.	$ 50
1201.—	Do	do	do	1	do	75
1202.—	Do	do	do	1¼	do	1 00
1203.—	Do	do	do	1½	do	1 25
1204.—	Do	do	do	1¾	do	1 50
1205.—	Do	do	do	2	do	2 00
1206.—	Do	Flat rim	do	¾	do	40
1207.—	Do	do	do	1	do	50
1208.—	Do	do	do	1¼	do	75
1209.—	Do	do	do	1½	do	1 00
1210.—	Do	do	do	1¾	do	1 25
1211.—	Do	do	do	2	do	1 50
1212.—	Do	Rounded rim,	2 double convex lenses	¾	do	75
1213.—	Do	do	do	1	do	1 25
1214.—	Do	do	do	1¼	do	1 75
1215.—	Do	do	do	1½	do	2 00
1216.—	Do	do	do	1¾	do	2 50
1217.—	Do	do	do	2	do	3 00
1218.—	Do	Flat rim	do	¾	do	65
1219.—	Do	do	do	1	do	1 00
1220.—	Do	do	do	1¼	do	1 25
1221.—	Do	do	do	1½	do	1 50
1222.—	Do	do	do	1¾	do	2 00
1223.—	Do	do	do	2	do	2 50
1224.—	Do	Bellows form,	1 double convex lens	¾	do	75
1225.—	Do	do	do	1	do	1 00
1226.—	Do	do	2 Double convex lenses	¾	do	1 00
1227.—	Do	do	do	1	do	1 25
1228.—	Do	do	do	1⅛	do	1 50
1229.—	Do	do	3 Double convex lenses	¾	do	1 50
1230.—	Do	do	do	1	do	1 75
1231.—	Do	do	do	1⅛	do	2 25
1232 —	Microscope on three legs with screw adjustment for focus.............					1 00
1233.—	Linen provers or microscope for counting threads in linen or wool fabrics. Hard rubber 1 inch open space..............................					2 00
1234.—	Do	do $\frac{18}{100}$ in. open circle..............................				75
1235.—	Do	do Brass, ¼ and ½ in. open space........................				75
1236.—	Do	do do $\frac{25}{100}$ in. square can be changed to $\frac{18}{100}$ in. Diameter for wool or linen fabrics..				75
1237.—	Reading glass, hard rubber frame,		1 double convex lens	1⅛	inch diam.	75
1238.—	Do	do	do	1⅜	do	1 00
1239.—	Do	do	do	1⅝	do	1 25
1240.—	Do	do	do	2	do	1 50
1241.—	Coddington lens, Brass frame $\frac{7}{16}$ in. diameter..........................					1 50
1242.—	Do	do ¾	do			2 00
1243.—	Do	do 1⅛	do			2 50
1244.—	Do	Silver frame..........................				2 50
1245.—	Do	do with cover..........................				3 50
1246.—	Do	do Large size with cover..................				5 50

READING AND PICTURE LENSES.

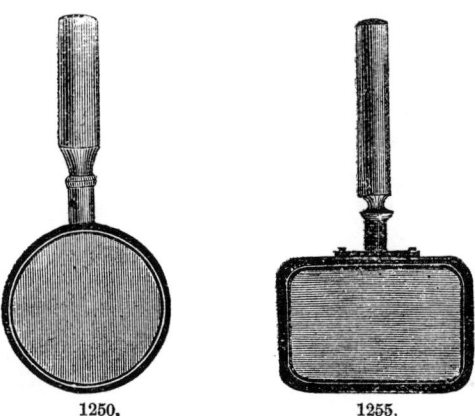

1250.   1255.

No.						Price·
1250.—Reading Glasses, Hard Rubber Frame, Double Convex lens, 2 in. diam.						$1 00
1251.—	Do	do	do	2½	do	1 50
1252.-	Do	do	do	3	do	2 00
1253.—	Do	do	do	3½	do	2 50
1254.—	Do	do	do	4	do	3 00
1255.—	Do	do	do	2 in. long		1 25
1256.—	Do	do	do	2½	do	1 75
1257.—	Do	do	do	3	do	2 25
1258.—	Do	do	do	3½	do	2 75
1259.—	Do	do	do	4	do	3 25

We are prepared to furnish Achromatic Microscopes, and accessories Manufactured by "Beck" London, the price of which are the same as those in London, U. S. duties and Freight charges only being added. Have also for sale, Queens School, College, Household, Popular, Students, Educational and Family microscopes, at prices varying from 6 to 100 dollars each. Materials, objects, Cabinet, and all implements connected with microscopes furnished at short notice.

ACHROMATIC MARINE, FIELD AND OPERA GLASSES.

These Glasses are designated and priced according to the diameter of the object glasses in French lines, as follows :

11 Lines, which is equal to 1 inch.
13 Do   do $1\frac{3}{16}$ inches.
15 Do   do $1\frac{5}{16}$ inches.
17 Do   do $1\frac{1}{2}$ inches.
19 Do   do $1\frac{11}{16}$ inches.
21 Do   do $1\frac{7}{8}$ inches.
24 Do   do $2\frac{1}{8}$ inches.
26 Do   do $2\frac{5}{16}$ inches.

MARINE AND FIELD GLASSES.

The power and sharpness of definition of an Opera or Field Glass depends upon the diameter of the object-glass; the greater the diameter the higher the power, and more clearly distant objects are seen.

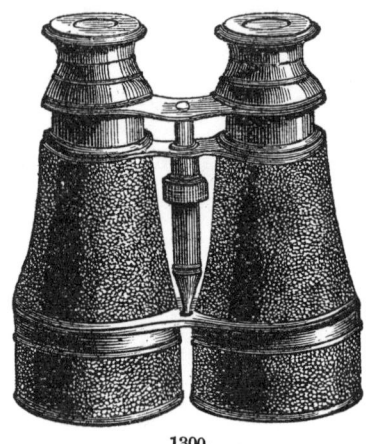

1300.

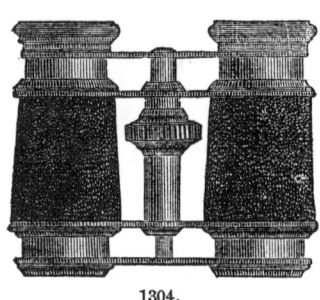

1304.

No.   Price.

1300.—U. S. Army Signal Service Six Lens Achromatic Marine or Field Glass, metal body, covered with Turkey morocco, sun shade to extend over the object-glasses, and heavy leather case, with strap.

Body 5⅜ inches long ; object-glasses 21 lines in diameter. $17 00
Do   5⅞   do   do   24   do 20 00
Do   6¼   do   do   26   do 22 00

1301.—Six Lens Achromatic Field Glass, metal body, covered with morocco, sun shades to extend over the object-glasses, and leather case with strap

Body 4¾ inches long, Object-glasses 21 lines in diameter.............. $10 50
Do   5¾   do   do   24   do 13 00
Do   6¼   do   do   26   do 15 00

1302.—Bardou's Six Lens Achromatic Marine or Field Glass, body covered with Turkey morocco, sun shades to extend over the object-glasses, in fine Leather case, with strap ; the best article made.

Body 6½ inches long, Object-glasses 26 lines in diameter................ 30 00
Do   do   do   do   do   with hinge adjustment for different widths of eyes 33 00

1303.—Six Lens Achromatic Glass, with three adjustable eye-pieces of different powers for Field, Marine, or Opera, metal bodies, covered with finest

No.   Price.

Turkey morocco, sun shades to extend over the object-glasses, and fine leather cases, with strap.

Body 3½ inches long, object-glasses 17 lines in diameter,............ $18 00
  "   4¼   "   "   19   " 22 00
  "   5   "   "   21   " 27 00
  "   5¾   "   "   24   "· 32 00

1304.—Six Lens Achromatic Field Glass, Rock Crystal Lenses, double adjustment of focus, so that, when closed, the instrument can be conveniently crrried in the pocket, in morocco case, without strap;   very powerful.

Body 2 inches long, object-glasses 10 lines in diameter............. 18 00
  "   2¾   "   "   11   " 20 00
  "   2¾   "   "   15   " 22 00

OPERA GLASSES.

1308 to 1313.

1305.—Opera Glasses, six lens achromatic, aluminium bodies, covered with fine black Turkey morocco; tubes and cross pieces japanned black; these are the lightest articles ever made.

Bodies 2½ inches long, object-glasses 15 lines in diameter, each...... $22 00
  "   3   "   "   17   "   " 27 00
  "   3½   "   "   19   "   " 31 00

1306.—Opera Glasses, six lens achromatic; bodies, tubes and cross pieces all black, bodies of a new and elegant shape, covered with finest Turkey morocco.

Body 2½ inches long, object-glasses 12 lines in diameter, each........ 10 50
  "   2⅝   "   "   13   "   " 11 00
  "   2¾   "   "   15   "   " 12 00

1307.—Opera Glasses, the same as 1306, but with gilt tubes, and bodies covered with fancy colored leather, and oxidized ornaments at top and base.

Body 2½ inches long, object-glasses 12 lines in diameter, each........ 12 00
  "   2⅝   "   "   13   "   " 13 00
  "   2¾   "   "   15   "   " 14 00

BARDOU'S OPERA GLASSES.

No.   Price.

1308.—Bardou's Opera Glasses, six lens achromatic; bodies, tubes and cross pieces all black; bodies covered with best Turkey morocco; cross pieces curved.

Body 2¼ inches long, object-glasses 13 lines in diameter, each					$8 50
" 2½ "	"	15	"	"	9 00
" 3 "	"	17	"	"	11 50
" 3¼ "	"	19	"	"	13 50

1309.—Bardou's Opera Glasses, the same as 1308, but with eight lenses.

Body 2¼ inches long, object-glasses 13 lines in diameter, each					10 50
" 2½ "	"	15	"	"	11 50
" 3 "	"	17	"	"	13 50
" 3¼ "	"	19	"	"	16 00

1310.—Bardou's Opera Glasses, the same as 1309, but with twelve lenses.

Body 2¼ inches long, object-glasses 13 lines in diameter, each					15 00
" 2½ "	"	15	"	"	16 00
" 3 "	"	17	"	"	18 00
" 3¼ "	"	19	"	"	20 00

1311.—Bardou's Opera Glasses, six lens, rock crystal, achromatic; bodies, tubes, and cross pieces all black; bodies covered with best Turkey morocco; cross pieces curved.

Body 2¼ inches long, object-glasses 13 lines in diameter, each					10 50
" 2½ "	"	15	"	"	11 00
" 3 "	"	17	"	"	13 00
" 3¼ "	"	19	"	"	16 00

1312.—Bardou's Opera Glasses, the same as 1311, but with eight lenses, rock crystal.

Body 2¼ inches long, object-glasses 13 lines in diameter, each					13 00
" 3 "	"	15	"	"	13 75
" 3½ "	"	17	"	"	16 00
" 4 "	"	19	"	"	18 00

1313—Bardou's Opera Glasses, the same as 1311, but with twelve lenses, rock crystal.

Body 2¼ inches long, object-glasses 13 lines in diameter, each					18 00
" 2½ "	"	15	"	"	19 00
" 3 "	"	17	"	"	20 00
" 3¼ "	"	19	"	"	22 00

PEARL GLASSES.

1314.—Opera Glasses, six lens achromatic; white pearl bodies, gilt tubes and cross pieces. raised eye-pieces.

Body 2⅝ inches long, object-glasses 13 lines in diameter, each					$16 00
" 3 "	"	15	"	"	17 00
" 3¼ "	"	17	"	"	19 00

In addition to the foregoing list of opera glasses, we are constantly receiving new and handsome designs, mounted in pearl, enamel, and oxidized metal.

ACHROMATIC TELESCOPES.

1325.

No.   Price.

1325.—Telescope, wood body, 3 draws, 15 inches drawn out, 6 inches shut, object-glass 1 inch in diameter, power 15 times $3 00

1326.—Telescope, wood body, 3 draws, 16 inches drawn out, 6 inches shut, object-glass 1⅛ inches in diameter, power 20 times..................... 4 00

1327.—Telescope, wood body, 3 draws, 23 inches drawn out, 8 inches shut, object-glass 1¾ inches in diameter, power 25 times................... 6 00

1328.—Telescope, wood body, 3 draws, 30 inches drawn out, 10 inches shut, object- glass 1⅝ inches in diameter, power 30 times.................. 8 00

1329.—Telescope, wood body, 4 draws, 37 inches drawn out, 11 inches shut, obect-glass 1⅞ inches in diameter ; superior glass ; power 35 times ... 14 00

1330.—Telescope, wood body, 4 draws, 42 inches drawn out, 11½ inches shut, object-glass 2⅛ inches in diameter, power 40 times 25 00

1331.—Telescope, wood body, 4 draws, 48 inches drawn out, 13⅛ inches shut, object-glass 2⅜ inches in diameter, power 50 times 36 50

1332.—Telescope, wood body, 5 draws, 28 inches drawn out, 7¾ inches shut, odject–glass 1⅝ inches in diameter ; about the same power as No. 1329, but more portable ; power 35 times 12 00

1333.—Telescope, wood body, 6 draws, 17 inches drawn out, 4¾ inches shut, object-glass 1⅛ inches in diameter, power 20 times 6 50

1334.—Telescope, wood body, 6 draws, 16 inches drawn out, 4¼ inches shut, object-glass ⅞ inch in diameter ; very portable : power 15 times...... 6 00

1335.—Telescope, brass body, covered with cord or leather ; shade to keep off the sun and rain ; 1 draw, 36 inches drawn out, 20 inches shut, power 25 times ; object-glass 1⅛ inches 13 00

1336.—Same as above, but with 2 or 3 draws, 15 inches shut.................. 13 00

1340.—Naval Achromatic Spy-glass, tapering wood body and one draw, 55 inches long when drawn out, 45 inches long when shut up ; rack and pinion for adjusting the focus. Power 50 times............................. 45 00

TOURISTS' GLASSES.

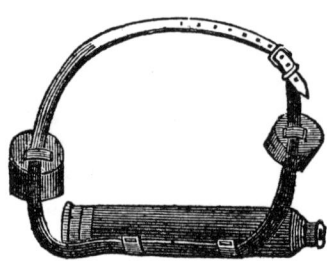

1341.

No.   Price.

1341.—Tourist's Achromatic Spy-glass, with brass body, covered with black
Turkey morocco ; three draws, 17 inches long when drawn out, 6 inches long when shut up ; object-glass 1¼ inches diameter ; sun shade
to slip beyond the object-glass ; heavy leather caps to cover both the
eye-glass and object-glass ; strong leather strap to sling over the
shoulder. Power 20 times $12 00

1342.—Same as No. 1341, but is 21 inches long when drawn out, 7 inches long
when shut up ; object-glass 1⅜ inches diameter. Power 25 times 15 50

1343.—Same as No. 1341, but is 24 inches long when drawn out, 9 inches long
when shut up ; object-glass 1⅝ inches diameter. Power 30 times 21 00

1344.—Same as No. 1341, but has four draws, and is 36 inches long when drawn
out, 10 inches long when shut up ; object-glass 1⅞ inches diameter.
Power 35 times 30 00

1345.—Wooden Tripod Stand, with vertical and horizontal motion, upon which
to place a spy-glass ; an exceedingly useful article, as a glass of much
power cannot be held in the hand with sufficient steadiness to pro-
duce the best effect 5 00

1346.—Brass Clamp with Gimlet Screw, to fasten a spy-glass to a post or tree,
three sizes to fit any of the foregoing spy-glasses .... 75, 1 25 and 1 75

DRAWING PAPER, COLORS, &c.

In filling orders for Instruments from different parts of the country, we have frequently been called upon to furnish, in addition, the various materials, for the office work of the Surveyor and Engineer, such as Drawing Paper, Colors, Text-Books, &c. We have therefore supplied ourselves with an assortment of these goods, and shall hereafter be able to furnish them on as favorable terms as any other dealer in the Union.

DRAWING STATIONERY.

WHATMAN'S HOT AND COLD PRESSED DRAWING PAPERS, SELECTED, BEST QUALITY.

No.				Price.
1400.—Demy,	15 × 20, per quire			$1 00
1401.—Medium,	17 × 22,	"		1 50
1402.—Royal,	19 × 24,	"		2 00
1403.—Super Royal,	19 × 27,	"		2 50
1404.—Elephant,	23 × 28,	"		3 50
1405.—Imperial,	22 × 30,	"		3 50
1406.—Columbier,	23 × 34,	"		5 25
1407.—Atlas,	26 × 34,	"		5 25
1408.—Double Elephant,	27 × 40,	"		6 00
1409.—Antiquarian,	31 × 53,	"		30 00

We only keep the best Whatman's Paper in stock, but to parties desiring it, can furnish the second quality at about 15 per cent. below the above prices.

DRAWING PAPER.

CONTINUOUS IN ROLLS.

Bleached Manilla, Buff Tint, for working Drawings, best American make, in Rolls of 50 to 100 lbs.

1415.—36 inches wide, thick, per pound, 15 cts						per yard $	10
1416.—40	"	"	"	"	15		"	12
1417.—44	"	"	"	"	15		"	14
1418.—48	"	"	"	"	15		"	16
1419.—54	"	"	"	"	15		"	18
1420.—58	"	"	"	"	15		"	22

WHITE ROLL GERMAN DRAWING PAPER.

Extra White, in Rolls, of 30 to 50 lbs.

No.						Price.
1425.—36 inch wide, medium, per pound, 40 cts...................... per yard $						25
1426.—44 " " " 40 .... "						30
1427.—56 " " " 40 .... "						35
1428.—56 " " extra tough, per pound, 55 cts......... "						50
1429.—56 " thick " " 55 .... "						70

BEST EGGSHELL DRAWING PAPER.

IN ROLLS OF 30 TO 40 POUNDS.

1430.—Eggshell 58 in. wide, thin, rough surface, per lb. 58 cts. .... per yard $						45
1431.— " 58 " medium " " 55 .... "						55
1432.— " 58 " thick " " 55 .... "						75
1433.— " 42 " medium " " 55 .... "						40

☞ Full Rolls, only of Continuous paper, sold by the pound, at above rates.

MOUNTED DRAWING PAPER.

WHITE, MOUNTED ON MUSLIN.

In Rolls of 10 yards.

1440.—German Paper, 42 inches wide, rough surface, per roll 9 00,.... per yard $1 00							
1441.— Do 54 do do do 12 00,.... do 1 40							
1442.— Do 54 do thick do do 14 00,.... do 1 60							
1445.— Eggshell, 42 do medium do do 12 50,.... do 1 45							
1446.— Do 54 do do do do 14 00,.... do 1 60							
1447.— Do 58 do do do do 16 00,.... do 1 80							

Large pieces for City, County or State Maps.—Mounted to order.

TRACING OR VELLUM CLOTH.

In Rolls of 24 yards, both sides glazed, or face glazed and back dull, suitable for pençil marks.

1450.—Imperial, 18 inches wide, per roll $6 50, per yard........................ $						35
1451.— Do. 30 do. do. 9 50, do.						50
1452.— Do. 36 do. do. 11 00, do.						60
1453.— Do. 42 do. do. 15 00, do.						75
1454.—Sagar's Patent, 18 do. do. 7 00, do.						35
1455.— Do. 30 do. do. 9 50, do.						50
1456.— Do. 36 do. do. 11 00, do.						60
1457.— Do. 42 do. do. 15 00, do.						75

THE NEW LINEN TRACING PAPER.

TRANSPARENT, VERY STRONG, AND WATERPROOF.

1460.—In Rolls of 20 yards, 36 inches wide, per roll $4.40, per yard........... $						35
1461.— Do. do. 48 do. do. 5.85, do. , ...						45

FRENCH TRACING PAPER.

FINE QUALITY, VERY CLEAR AND STRONG.

No.				Price.
1465.—In Sheets, Royal,	19×25 inches, per quire......................			$1 00
1466.— " Super Royal,	21×26 "	"	....	1 50
1467.— " Double Elephant,	28×40 "	"	... .	2 50
1468.—In Rolls, 11 yards long and 48 inches wide, per roll,...................				1 50
1469.— " 22 "	"	"	"	2 50
1470.—Vegetable Royal,	19×25 inches, per quire $2 20, Per sheet.....			15
1471.— " Super Royal,	21×26 "	" 3 50,	"	40
1472.— " Double Elephant,	28×40 "	" 10 00,	"	65
1473.— " "	in rolls of 22 yards, 54 inches wide, per roll,..			5 00

PROFILE PAPER.

1475.—Plate A, 42 × 15in., horizontal ruling, 4, vertical, 20 to in., per sheet								$ 40
1476.—Plate B, 42 × 13¼	"	4,	"	30	"	"	.....	40
1477.—Plate C, 42 × 15	"	5,	"	25	"	"	.....	40
Nos. 1475, 1476, and 1477, per quire....................................								8 50
1479.—Continuous Profile Paper Plates, A or B, 22 inches wide, per yard.......								30

MUSLIN BACKED ROLL PROFILE PAPER.

1480.—Muslin Backed Roll Profile Paper, of either Plate A or B, 22 inches wide, in rolls of 20 yards, per yard .... $ 75

1481.—Muslin Backed Roll Profile Paper, Plate B, 9 inches wide, in rolls of 20 yards, per yard... 50

Plate B corresponds to that in sheets known as Brown's Profile Paper.

CROSS SECTION PAPERS.

1482.—Topographical Paper, 14 × 17 inches, ruled 400 feet to the inch, per sheet, 10 cents.................................. ....per quire $1 75

1483.—Trautwine's Cross Section and Diagram, 10 feet to inch, for embankments of 14 and 24 feet, roadway, and for excavations of 18 and 28 feet, rulings 19¾ × 12 inches, per sheet. 25c......................per quire 5 00

1484.—Cross Section Papers, rulings 22×16 in. 8 ft. to in. per sheet, 25c.,					"		5 00
1485.— "	"	20×16	10	"	"	25c.,	"	5 00
1486.— "	"	20×16	10	"	"	every fifth line		5 00

heavy, per sheet 25c., per quire... 5 00

1487.—Cross Section Papers, rulings 22×16 inches, 16 feet to inch, per sheet 25c. 5 00

All the Profile and Cross Section Papers can be furnished, printed with red or green lines.

LYONS' TABLES.

No.

1490.—Lyons' Tables. A set of Tables for finding at a glance the true cubical contents of Excavation and Embankments for all Bases, and for every variety of Ground and Side Slopes. By M. E. Lyons, C. E.

Sheet No.	1.	General Table for all Bases and all Slopes.			
Do	2.	For Side Hill Cuts and Fills.			
Do	3.	Base	12 feet	Slopes	1½ to 1
Do	4.	do	14	do	1½ to 1
Do	5.	do	15	do	¼ to 1
Do	6.	do	15	do	1 to 1
Do	7.	do	15	do	1½ to 1
Do	8.	do	16	do	¼ to 1
Do	9.	do	16	do	1 to 1
Do	10.	do	18	do	¼ to 1
Do	11.	do	18	do	¾ to 1
Do	12.	do	18	do	1 to 1
Do	13.	do	18	do	1½ to 1
Do	14.	do	20	do	1½ to 1
Do	15,	do	24	do	¼ to 1
Do	16.	do	24	do	1½ to 1
Do	17.	do	25	do	1½ to 1
Do	18.	do	26	do	1½ to 1
Do	19.	do	28	do	¼ to 1
Do	20,	do	30	do	1 to 1
Do	21.	do	30	do	1¼ to 1
Do	22.	do	30	do	1½ to 1
Do	23.	do	32	do	1 to 1
Do	24.	do	32	do	1½ to 1

The Tables are printed in clear, bold type, on tinted paper, sheets 25 × 16 inches. They may be used by candle-light without injuring the eyesight. Each sheet is complete in itself, and embraces all that is wanted in connection with Base or Slope designated, whether on level or side-hill cross section.

Per sheet, 25 cents ; bound in one volume. $8 50

FIELD BOOKS.

1491.—Level Books, $7 00 per dozen .... each $ 70

1492.—Transit Books, 7 00 do .... do 70

1493.—Record Books, 7 00 do .... do 70

1494.—Cross Section Books, 8 inches long by 7 inches wide, for Topography, $10 00 per dozen .... each 1 00

BOUND PROFILE BOOKS.

These books are for field or office purposes, being printed on both sides of a tough thick paper, and bound in flexible covers, convenient for the pocket. Each page will contain a profile of three thousand feet in length, so that each folio will contain an average section of a road as usually laid

out for construction. Railroad and other engineers will find them very
useful. Size of book 9½ by 5¾ inches. The rulings correspond to our
large profile plates A and B.

No.							Price.
14.5.—Plate A, 25 leaves imitation Turkey morocco, with elastic band						....	$3 50
1496.—	"	50	"	"	"	"	5 00
1497.—	"	100	"	"	"	"	8 00
1498.—	"	50	"	Turkey morocco, turned edges,	"	....	6 00
1499.—	"	100	"	"	"	"	9 00
1500.—Plate B, 25	"	imitation Turkey morocco,	"	....	3 50
1501.—	"	50	"	"	"	"	5 00
1502.—	"	100	"	"	"	"	8 00
1503.—	"	50	"	Turkey morocco, turned edges,	"	....	6 00
1504.—	"	100	"	"	"	"	9 00

INK SLABS, SAUCERS AND WATER COLORS.

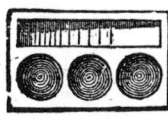

1511.   1515.

Ink Slabs.

For India Ink and Colors; containing 3 holes or cups and 1 slanting division.

1511.—Measuring 2¾ × 1½ inches, each	....	$	15
1512.—	"	3¾ × 2⅜	"	....	25
1513.—	"	4⅜ × 2¾	"	....	35
1514.—	"	4¾ × 3	"	....	40

Cabinet Nests, Water Glasses, and Palettes.

Porcelain Saucers in nests; fitted on each other.

1515.—Containing 5 Saucers and a Cover, 2½ inches in diameter, per nest.....							75
1516.—	"	5	"	" 2¾	"	"	"	85
1517.—	"	5	"	" 3¼	"	"	" 1 00
1518.—	"	5	"	" 3¾	"	"	" 1 25
1519.—Patent Ink Slab, 1¾ × 4½ inches, with Cover, each	....	60
1520.—	"	2⅛ × 5¼	"	"	....	75
1521.—Artist's Water Glass, 2¾ inches in diameter, each	....	15
1522.—	"	"	3¼	"	"	....	25
1523.—	"	"	3¾	"	"	....	35
1524.—	"	"	4¼	"	"	....	45
1525.—China Palettes, oval or oblong, 4 inches, each	....	30
1526.—	"	"	5	"	....	35
1527.—	"	oval or oblong, 6 inches, each	....	40
1528.—	"	"	7	"	....	45
1529.—	"	"	8	"	....	50
1530.—	"	"	9	"	....	60
1531.—	"	"	10	"	....	75

WINSOR & NEWTON'S CAKE WATER COLORS.

IN HALF OR WHOLE CAKES.

Whole Cake.

Half Cake.

No.   Price

1535.—Whole Cake 30 cents...Half Cake $ 15

1 Antwerp Blue	16 Flake White	31 Orange Chrome
2 Bistre	17 Gamboge	32 Payne's Grey
3 Blue Black	18 Hooker's Green, No. 1	33 Prussian Blue
4 British Ink	19 Hooker's Green, No. 2	34 Prussian Green
5 Bronze	20 Indian Red	35 Raw Sienna
6 Brown Ochre	21 Indigo	36 Raw Umber
7 Brown Pink	22 Italian Pink	37 Roman Ochre
8 Burnt Sienna	23 Ivory Black	38 Sap Green
9 Burnt Umber	24 King's Yellow	39 Terre Verte
10 Chinese White	25 Lamp Black	40 Vandyke Brown
11 Chrome Yellow	26 Light Red	41 Venetian Red
12 Cologne Earth	27 Naples Yellow	42 Vermillion
13 Deep Chrome	28 Neutral Tint	43 Yellow Lake
14 Dragon's Blood	29 New Blue	44 Yellow Ochre
15 Emerald Green	30 Olive Green	

1536.—Whole Cakes, 60 cents each................................Half Cakes   30

45 Black Lead	50 Indian Yellow	55 Rubens' Madder
46 Brown Madder	51 Mars Yellow	56 Scarlet Lake
47 Cerulean Blue	52 Neutral Orange	57 Scarlet Vermillion
48 Constant White	53 Purple Lake	58 Sepia
49 Crimson Lake	54 Roman Sepia	59 Warm Sepia

1537.—Whole Cakes, 85 cents each................................Half Cakes   45

60 Cobalt Blue	61 Orange Vermillion	62 Violet Carmine

1538.—Whole Cakes, $1.15 each...................................Half Cakes   60

63 Aureolin	69 French Blue (or	74 Lemon Yellow
64 Burnt Carmine	French Ultramarine)	75 Pink Madder
65 Cadmium Yellow, Pale	70 Gallstone	76 Pure Scarlet
66 Cadmium Yellow	71 Green Oxide Chro-	77 Rose Madder
67 Cadmium Orange	72 Indian Purple [mium	(or Madder Lake)
68 Carmine	73 Intense Blue	78 Viridian

1539.—Whole Cakes, $1.80 each...................................Half Cakes   90

79 Field's Orange Ver.	81 Mars Orange	83 Smalt
80 Madder Carmine	82 Purple Madder	84 Ultramarine Ash

1540.—Quarter Cake,.. each 2 25

85 Genuine Ultramarine.

WATER COLOR SLIDE-LID BOXES.

No.   Price.

1541.—Color Boxes to hold 6 whole or half cakes........................... . $ 40
1542.—   "   "   12   "   "   ....·   60
1543.—   "   "   18   "   "   80
1544.—   "   "   24   "   " 1 00

WINSOR & NEWTON'S WATER COLOR LIQUIDS.

IN GLASS BOTTLES.

1545.—Carmine.... .... $ 60 | 1549.—Extract of Ox Gall..... $ 50
1546.—Indelible Brown Ink...   60 | 1550.—Indian Ink............   50
1547.—Prout's Brown.........   60 | 1551.—Chinese White.........   50
1548.—Gold Ink..............   50 | 1552.—Pure Gold in shells   20
1553.—Pure Gold in cakes, $2.50; in cups...................................   25

WINSOR & NEWTON'S FULL CAKE WATER COLOR BOXES, FITTED.

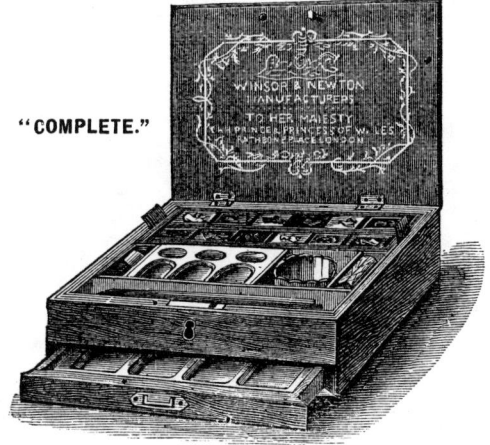

"COMPLETE."

1557.—" COMPLETE " BOX.

1555.—12 Cakes, polished Mahogany Lock and Drawer Box................... $7 50
1556.—18 do   do   do 11 00
1557.—12 do   do   Complete Box, fitted,................... 11 00
1558.—18 do   do   do   do 15 00
1559.—24 do   do   do   do 19 00

Half Cake Water Color Boxes, fitted.

1560.—12 Cakes, polished Mahogany Lock and Drawer Box................... 6 00
1561.— 18 do   do   do 8 00
1562.—12 do   do   Complete Box, fitted, ..., ... 7 00
1563.—18 do   do   do   do 9 00

CHINESE OR INDIAN INK.

This Ink is best adapted for shading, and is indispensable to Artists, for its brilliancy of shade; it is also preferable to any other ink for tracing purposes.

FULL SIZE.

1571.

1570.

1569.

1568.

1567.

1566.

No.		Price.
1565.—Oval, black, with Lion Head.....................................Cake.	$	25
1566.— " " "	"	50
1567.—Round, gilt, "	"	75
1568.—Square, black, gilt Figures	"	75
1569.—Oblong, " "	"	1 50
1570.— " gilt fine...	"	1 50
1571.—Square, black, gilt Figures (Winsor & Newton's best)............	"	2 00

JAPANESE INK.

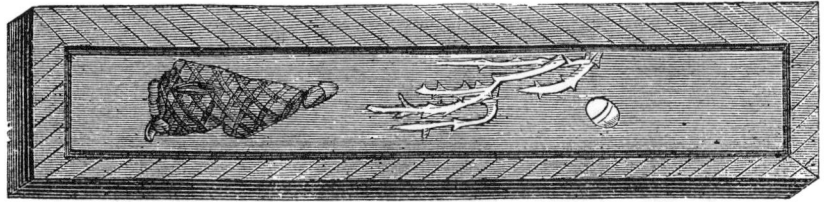

1573.—Full Size.

This Ink is best adapted for mechanical drawings, where lines are frequently washed, in applying brush tints. The lines drawn with this Ink, will remain clear and distinct, and will not be blurred or defaced by the brush.

1572.—Oblong, black, with Figures, best small cake..................per cake $1 00
1573.— " " " " large " .... " 3 00

These Inks are imported for us from China and Japan.

BRUSHES.

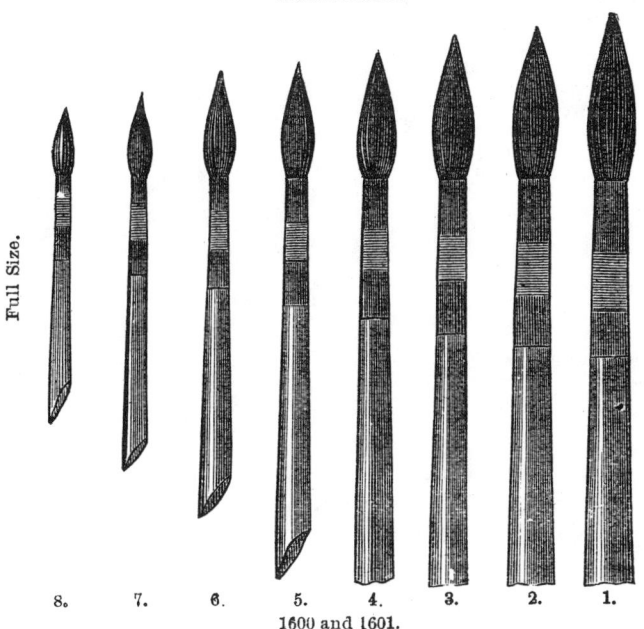

8.   7.   6.   5.   4.   3.   2.   1.

1600 and 1601.

No.

1600.—Red Sable in Quills

No. 1	2.	3.	4.	5.	6.	7.	8.
each $ 40	30	25	20	18	15	12	10

1601.—Camel Hair in Quills, No. 1 to 8 (fine quality)................each 5 to 10 cts.

Full Size.

1604 and 1605.

1604.—Red Sable in Swan Quills

No. 0.	1.	2.	3.	4.	5.	6.
each $ 2 00	1 50	1 20	1 00	75	60	50

1605.—Camel Hair in Swan Quills

No. 0.	1.	2.	3.	4.	5.	6.
each $ 75	50	40	30	20	18	15

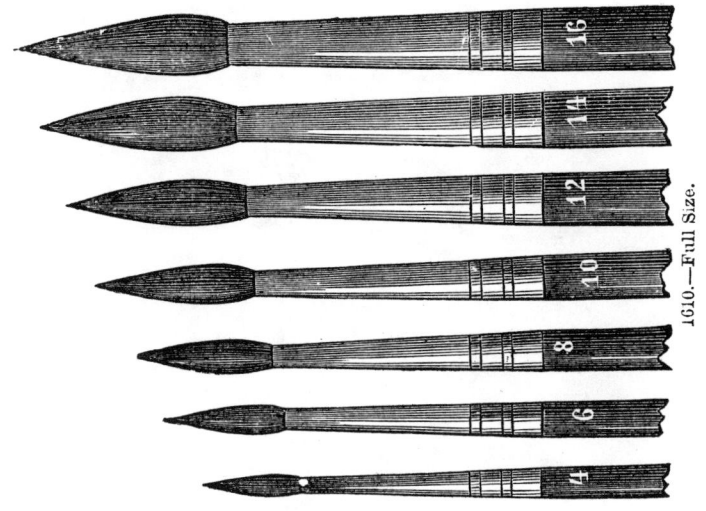

1610.—Full Size.

No.
1610.—Red Sable in Albata, with handle

No. 2.	4.	6.	8.	10.	12.	14.	16.	18.	20.	22.
each $ 25	25	30	40	50	60	75	1 00	1 50	1 75	2 00

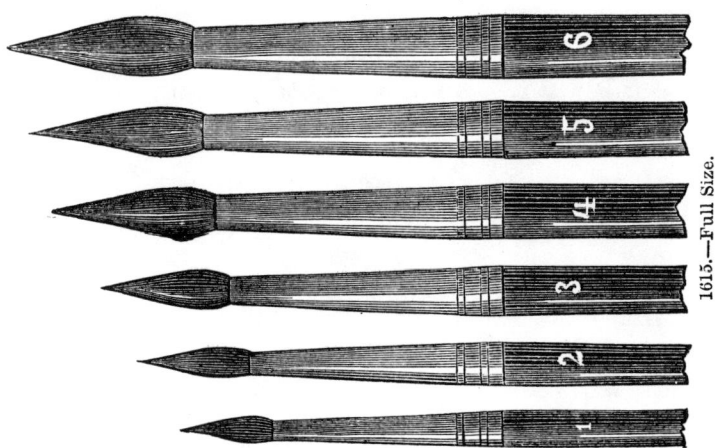

1615.—Full Size.

1615.—Camel Hair in Tin, with handle

No. 1.	2.	3.	4.	5.	6.
each $ 10	10	10	12	12	15

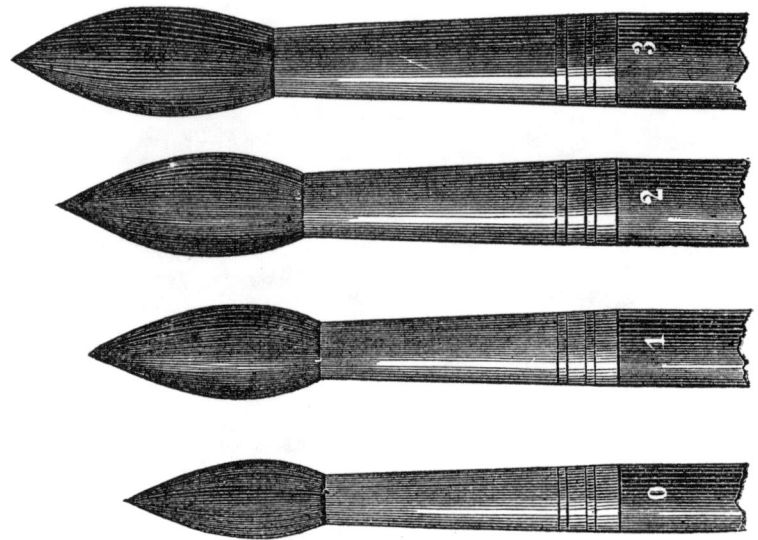

1620.—Full Size.

1620.—Camel Hair Sky or Wash Brush

No. 0.   I.   2.   3.

each $ 25   30   35   40

1625.—Half Size.

1623.—Camel Hair Wash Brushes in Tin, with 2 points, or double
sizes about equal to No. 1620

No. 0   I.   2.   3.

each $ 4v   50   60   75

1625.—Red Sable in Albata, with 2 points

No. 1.   2.

each $ 1 00   1 25

GILLOTT'S STEEL PENS.

No.		Price.
1690.—Mapping, on cards	per doz. $	75
1691.—Lithograph, on cards	"	75
1692.—Lithograph Crow Quil, on cards	"	75
1693.—Extra Fine No 303,	per gross	1 50
1694.— " " 170	"	1 00
1695.—Falcon Pens	"	75
1696.—Commercial Pens	"	75
1697.—Business Pens	"	75

QUILL PENS.

THREE SIZES.

No.				Price.
1698.—No. 1, per box of 2 doz.				$ 40.
1699.—No. 2,	"	"		60
1700.—No. 3,	"	"		75

LEAD PENCILS.

A. W. FABER'S.

1701.—Hexagon, very best Siberian, No. 4 B to 6 H			doz.	$1 50
1702.— " " Drawing, No. 1 to 5			"	1 00
1703.—Black round, best, No 1 to 4			"	75
1704.—Hexagon, for Divider Points, No. 4			"	1 00
1705.—Round, " " "			"	75

Artist Pencil with Siberian Lead.

1706.

1706.—Artist Pencil with Siberian lead, Large			each	35
1707.— " " " Small			each	25
1708.—Leads for Artist Pencils, Siberian, 6 in box			box	75
1709.— " best " 1 to 5			"	50

These leads fit the new pencil holders in Altender and Swiss sets.

1710.—Hexagon carmine and blue pencils, Faber, extra large			doz.	1 80
1711.—Round carmine, tipped			"	1 50
1712.— " blue, tipped			"	1 25
1713.— " green, tipped			"	1 25
1715.—One box, containing 7 Pencils, for Engineers, sliding ends to box				60
1716.— " " 5 " BB to H				60
1717.— " " 7 " BBB to HH				75
1718.— " " 10 " BBBB to HHHH				1 00

DRAWING PENCILS.

A. A.

1720.—Standard—Hexagon, purple polished, 6B to 6H ....per doz. $1 25

Polygrades, Nos. 1, 2, 3, 4, 5.

Best quality, in stained Cedar.

1721.—Hexagon, red polished, gilt, extra fine ....per doz. $ 75
1722.—Round, black polished, gilt, extra fine .... " 60

This grade we compare with the Imported Pencils known as Hexagon and Round Gilt.

1725.—Red Chalk Pencils for marking stake ....per doz. 50
1726.— " in lump ....per lb. 20
1727.—French s Venetian Crayons, for marking stakes (superior quality) per doz. 60

SPONGE RUBBER.

FOR CLEANING DRAWINGS.

No.		Price.
1730.	Sponge Rubber, small cakes	each $ 10
1731.	Do medium "	" 30
1732.	Do large "	" 75
1733.	Sponge Rubber and Eraser	" 20

INDIA RUBBER.

1735.	A. W. Faber's First Quality, white, $1\frac{1}{2} \times 1$ inch	each $ 6
1736.	Do do do $1\frac{3}{4} \times 1\frac{1}{8}$ "	" 8
1737.	Do do do $1\frac{3}{4} \times 1\frac{1}{4}$ "	" 10
1738.	Do do do $2 \times 1\frac{3}{8}$ "	" 12
1739.	Do do do $2 \times 1\frac{3}{8}$ " thick	" 15
1740.	Do do do $2\frac{1}{4} \times 1\frac{1}{2}$ "	" 20
1741.	Do do do $3 \times 2\frac{1}{8}$ "	" 40
1742.	Do do Black pure Gum, $1\frac{3}{8} \times \frac{13}{16}$ inch	" 6
1743	Do do do $1\frac{5}{8} \times 1\frac{1}{3}$ "	" 10
1744.	Do do do $1\frac{7}{8} \times 1\frac{1}{4}$ "	" 15
1745.	Do do do $1\frac{7}{8} \times 1\frac{1}{8}$ " thick	" 18
1746.	Do Improved Ink Eraser, $1\frac{1}{2} \times 1$ inch	" 6
1747.	Do Combined Ink and Pencil Eraser	" 20
1748.	Do do do do Mammoth	" 40

STEEL ERASERS.

1750.	Rogers & Son's Steel Blade, Cocoa Handle	$ 50
1751.	" " Bone "	60
1752.	" " Long Ivory Handle	75
1753.	" Knife Eraser, Ebony "	75
1754.	" " Ivory "	1 25

DRAWING BOARDS.

1755.	R. P. I., 14×10 inches	$ 75
1756.	Super Royal size, dovetailed ends	1 50
1757.	Double Elephant size, "	3 00
1759.	Framed Drawing Board, Mahogany, centre soft Pine and removable, 17×12 inches	3 00
1760.	Do do do 27×19 inches	5 00

MISCELLANEOUS.

No.							Price·
1775.—Best Foolscap Paper, per ream							$5 50
1776.—Best Letter Paper, "							5 00
1777.—Best Commercial Note, "							3 50
1778.—Superior Post Office Paper, buff tint, 19 × 24, per ream							7 00
1780.—Specification Paper, extra fine Legal Cap, thick, per ream							5 75
1781.—Flat Paper, smooth, extra, 16 × 21, per ream							9 50
1782.— " " " 18 × 23, "							13 50
1783.—Superior White Envelopes, per thousand							4 00
1784.— " Buff " "							3 50
1785.— " Buff " "legal" (large size), per thousand							8 75
1786.—Arnold's Writing Fluid, per quart							75
1787.—Maynard and Noyes' Writing Ink, per quart							75
1788.—Blue Ink, per bottle							25
1789.—David's Carmine, 2 ounce bottles, with glass stoppers, per bottle							50
1790.—Rubber Bands, ¼ inch wide, 2 inches long, per gross $1 15, per doz							12
1791.— " ¼ " 2½ " " 1 40, "							15
1792.— " ¼ " 3 " " 1 65, "							20
1793.— " ¼ " 3½ " " 2 00, "							25
1794.— " ½ " 2 " " 2 25, "							25
1795.— " ½ " 2½ " " 2 50, "							30
1796.— " ½ " 3 " " 2 75, "							35
1797.— " ½ " 3½ " " 3 00, "							40
1798.— " ½ " assorted length " 2 00,							
1799.— " 1⁄16 " 1¼ inches long for tickets, &c., per gross.							25

All other sizes Rubber Bands furnished at proportional rates.

1800.—Mucilage, per cone (3 oz.)		25
1801.— " " quart		1 25

Copying Books, Copying Ink and Presses, Blotting Paper, and all articles of Stationery needed in Engineers' offices furnished at reasonable rates.

Envelopes, Letter and Note Heads, Cards, &c., printed and lithographed at usual prices.

1802.—Timber Scribes or Marking Irons, each		1 25
1803.—Blow Pipes, Brass, plain		50
1804.— " " with bulb		75
1805.—Geological Hammers, each		75
1806.—Steel Chisels, 5 inches long, each		25
1807.— " 6 " "		30
1808.—Chamois Skins		75

Arkansas Oil Stones, 25c. to $2 00.

Thermometers, all kinds and sizes, 40c. to $10 00.

Barometers (Aneroid and Mercurial.)

Hydrometers, for testing Spirits, Ammonia, Ether, Alkalies, Vinegar, Molasses, Salt Water, Urine, Milk, Oils, Beer, Bark, &c., 90c. to $2 00.

Twaddels, 1 to 5, $1 25 to $1 75.

TURNING LATHE CHUCKS.

Horton Chucks.

1810.—	4 in.	6 in.	9 in.	12 in.	15 in.	18 in.
	each $26 00.	26 00.	34 00.	44 00.	52 00.	62 00.

Washburn & Whiton Chucks.

1811.—	3 in.	4 in.	5 in.	6 in.	9 in.
	each $12 00.	15 00.	16 00.	18 00.	26 00.

Oneida Community Chucks.

1812.—	3 inch.	6 inch.
	each $11 00.	26 00.

Beach Patent Drill Chuck.

No.   Price

1813.—No. 0. holds from 0 to ⅛ inch. Jewellers.............................. $8 00
1814.— " 1.   "   0 to ¼ "   diameter....................... 8 00
1815.— " 2.   "   0 to ⅜ "   " 8 50
1816.— " 3.   "   0 to ½ "   " 10 00
1817.— " 4.   "   1/16 to ⅝ "   " 11 00

Morse Twist Drills supplied at Manufacturers' prices,
Stub's Steel and Steel Wire.

RAIN GAUGE.

1820.—Smithsonian Rain Gauge, made entirely of brass. This gauge has been
adopted by the Smithsonian Institute and U. S. Patent Office, and is
the most simple in its construction of any now in use. It is furnished
with a graduated scale which reads to 10ths and 100ths of inches: also
a wooden cylinder to insert in the ground for the protection and ready
adjustment of the instrument.. $5 00

DEMONSTRATION LENSES.

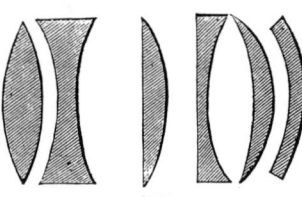

1821.

1821.—Demonstration Lenses, a set of six; 1¾ inches diameter, showing the
forms of the various kinds of lenses, viz: Double-Convex, Double-
Concave, Plano-Convex, Plano-Concave, Meniscus-Convex, and Meniscus-Con-
cave, per set.. $2 50

COSMORAMA LENSES.

No.   Price.

1822.—Double or Plano-Convex Lens, 8 inches diameter, and either 30, 36, 48
or 72 inches focus, each............................. .... $5 00

1823—Double or Plano-Convex Lens, 7 inches diameter, same foci as 1822, each 4 00

1824.—Double or Plano-Convex Lens, 6 inches diameter, of either 24, 30, 36, 48,
or 72 inches focus, each... 3 00

1825.—Double or Plano-Convex Lens, 5 inches diameter, of either 18, 20, 24, 30,
36, 48 or 72 inches focus, each... 2 50

1826.—Double or Plano-Convex Lens, 4 inches diameter, of either 12, 14, 16, 18,
20, 24, 30, 36, 48 or 72 inches focus, each............................. 1 50

1827.—Double or Plano-Convex Lens, 3 inches diameter, any focus 6 to 36 inches, each.. 1 00

1828.—Double or Plano-Convex Lens, 2 inches diameter, any focus 6 to 36 inches, each... 75

1829.—Double or Plano-Convex Lens, 1¼ inches diameter, any focus 5 to 48 inches, each... 50

PRISMS.

1831.—Solid Flint Glass Prisms, 3 inches long, each..............................					$ 65
1832.—	"	"	4	"	" 75
1833.—	"	"	5	"	" 1 00
1834.—	"	"	6	"	" 1 25
1835.—	"	"	7	"	" 1 50
1836.—	"	"	8	"	" 2 00

1837.—Metal Stand for Prisms, each $1.50 to................................. 3 00

1838.—Prisms for Stereoscopes, 1⅝ inches square, per pair.................... 75

1839.—Polyprism, making many heads out of one........................... 25

1840.—Set of two Prisms to illustrate the principle of the Achromatic Object
Glass... .... 3 00

ACHROMATIC OBJECT GLASSES.

FOR SPY-GLASSES AND TELESCOPES.

Achromatic Lenses are formed by a combination of a Double-Convex lens
of crown glass, and a Plano-Concave lens of flint glass. The advantages
of a lens formed in this manner are, freedom from spherical observation
or distortion, and the rays of light are not decomposed into the primary
colors, in other words, the light passes through the lens and suffers no
change thereby.

1845.—Acromatic Object Glass, 1½ inches diameter, 18 to 30 inches focus......								2 00
1846.—	"	"	1¾	"	18 to 30	"	......	3 50
1847.—	"	"	2	"	18 to 30	"	......	4 50
1848.—	"	"	Extra fine, 2 inches diameter, 36 inches focus..					7 00
1849.—	"	"	"	2½	"	44	"	.. 13 00
1850.—	"	"	"	3	"	48	"	.. 37 00
1851.—	"	"	"	3½	"	54	"	.. 50 00
1852.—	"	"	"	4	"	60	"	.. 90 00

STENCIL PLATES.

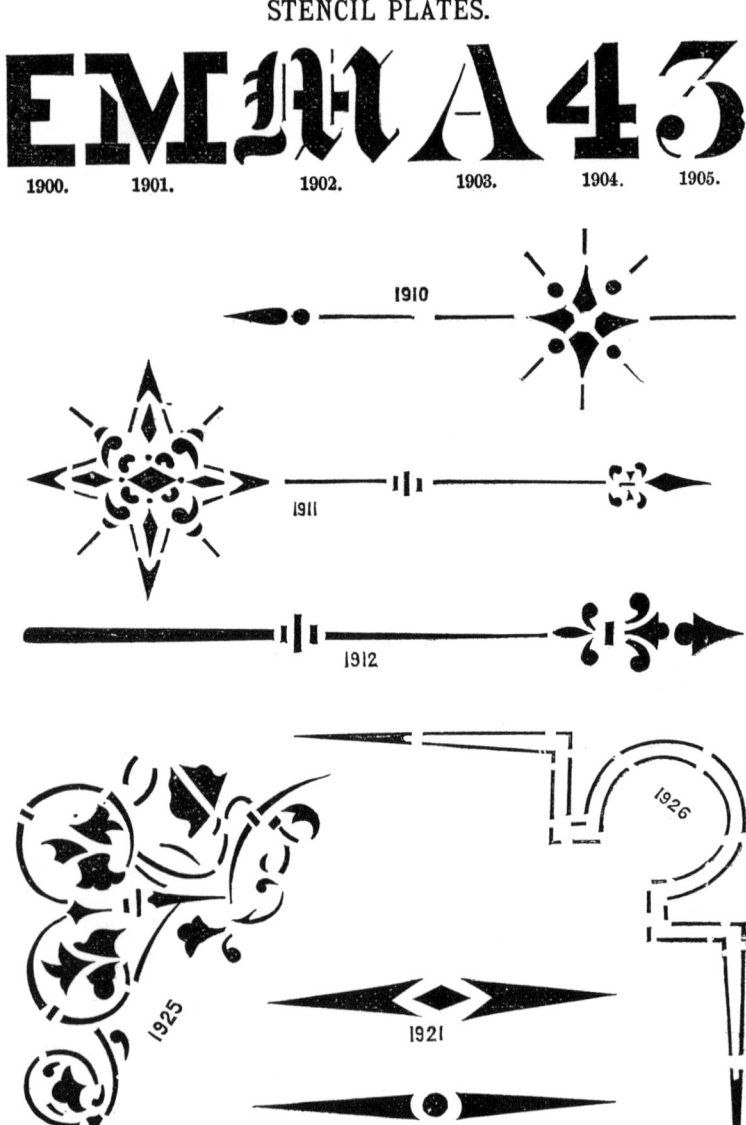

1900.   1901.   1902.   1903.   1904.   1905.

STENCIL PLATES.

HEIGHT OF LETTERS.	¼ in.	⅜ in.	½ in.	⅝ in.	¾ in.	1 in.
1900.—Alphabet..........................	2 00	2 00	2 25	2 25	2 25	2 45
Single letters, or words...per letter	8	8	9	9	9	10
1901.—Alphabet..........................	2 25	2 25	2 30	2 30	2 30	2 50
Single letters, or words...per letter	9	9	10	10	10	11
1902.—Alphabet.........................		2 25	2 35	2 35	2 50	2 60
Single letters, or words...per letter		10	10	10	11	11
1903.—Alphabet..........................	2 25	2 25	2 30	2 30	2 30	2 50
Single letters, or words...per letter	9	9	10	10	10	11

Figures.

Set of figures to match any of above styles of letters, the same price as a dozen letters.

No.		Price.
1910.—North point, full size...	$	50
1911.— " "		75
1912.— " "		50
1920.—Dasher, 1¾ inches long..		15
1921.— " 1½ "		15

(3 cents per inch for each additional inch.)

1925.—Ornamental Corner...		75
1926.— " "		75
Stencil Brushes, two sizes......................................8 cts and		12

We furnish to parties who order, 1 doz. numbers or letters, a stencil brush of each size, without extra charge.

These Stencil Plates will be found very useful to the architect, engineer, or surveyor, especially where words are repeated many times, as, Plan, Name of Railroad, Ground Plan, Elevations, Section, Profile, Drawing, No. — , etc.

USE INDIAN INK, THICK, FOR MARKING.

ANY WORD OR WORDS, AND LARGER LETTERS, CUT TO ORDER.

ERRATA.

Page 108, line 19, to read, " The moment the image ceases to run below is of course apparent."

  "   155, No. 277, to read " 377."

  "   164, No. 503½, " Protractor, 4 inch diameter, half circle, whole degree, centre on inner edge, $2.25."

  "   164, No. 504, price " $2.75."

  "   169, No. 602, price " $3.00."

  "   185, No. 1014, price " 50c."

W. & L. E. GURLEY'S

BOOK CIRCULAR,

1874.

Architecture, Bridges, Civil Engineering,
Surveying, Construction, Strength
of Materials, Geology, Mining,
Pocket Books, Tables,
Instruments, etc.

———— ◄❦► ————

*** NOTE.—Parties ordering should either send drafts on New York, or postal orders on Troy, N. Y.; or if money is enclosed in letters, such letters should be registered at the post-office where mailed.

We prepay postage on all Books when the price is sent to us in advance.

Orders for over $10.00, will be sent by express, " C. O. D.;" but for smaller sums parties will please remit the necessary amount with their order.

Write all letters legibly, give your post-office, county and State, and be sure to sign your letter before mailing.

We are not responsible for loss of goods sent by mail.

Should any other works on kindred topics be desired, we will furnish them at publishers' prices.

TROY, N. Y.:

PUBLISHED BY W. & L. E. GURLEY.

Table of Contents.

		PAGE.
Adjustment of Compasses		22, 23
Do	Vernier Transits	44–48
Do	Surveyors' Transits	66–68
Do	Solar Compass	82–85
Do	Solar Attachment	105, 106
Do	Y Leveling Instruments	120, 121, 124, 125, 126, 127, 128, 129, 130
Do	Builders', or Dumpy Level	132
Adjusting Socket		86
Do	price of	87
Alteneder's Patent Joint Drawing Instruments		152–154
Aneroid Barometers		7, 8
Architects' Ivory Scales		169
Do	Boxwood Scales	170
Attachments of Telescope		41, 42
Barometers, Aneroid		7, 8
Ball Spindle		29
Books		219–234
Boston Rod		134
Boxes for Colors		205
Bronze Finish		140
Brass Drawing Instruments		154–157
Brushes		207–210
Builders' Level		131, 132
Centrolinead		185
Chains, American		8, 141, 142, 143
Do	Spanish or Mexican	8, 9, 141, 144
Do	Grumman's Patent	9, 143, 144
Do	French or Metre	9
Chain Scale, Ivory		169
Do	Boxwood	170
Clamp and Tangent		42
Cosmorama Lenses		215
Compasses, Plain		5, 20–28
Do	Vernier	5, 29–33
Do	Railroad	5, 58–62
Do	Extras to,	5
Do	Solar	5, 70–101
Do	Pocket	6, 137, 138
Do	Miners'	6, 136, 137
Compound Ball		50

TABLE OF CONTENTS.

	PAGE
Cross Wire and Ring	36
Cross Section Paper	201
Curves, Wood	181, 182
Do Rubber	183
Demonstration Lenses	214
Diurnal Variation	57
Drawing Instruments, Swiss	146–152
Do Alteneder's Patent Joint	152-154
Do Brass	154-157
Do German Silver	157-163
Drawing Paper	199–201
Do Boards	212
Dumpy Level (*See Builders' Level.*)	
Equation of Time	102
Erasers	212
Eye piece, how composed	35
Field Glasses	193–195
Field Books	202
Flag Staff	7
German Silver Drawing Instruments	157–163
General Matters	139
Grumman's Patent Chain	9, 143, 144
Hand Level	7, 187, 188
Hyperbolas	179
Information to Purchasers	13–15
India Ink	206, 207
Jacob Staff Mountings	21
Do Socket	24
Lathe Chucks	214
Lacquering	140
Leveling Instruments	7
Do Rods	7, 133, 134, 135, 136
Do Instruments, Y Levels	119–130
Do Instruments, Dumpy, or Builders'	131, 132
Limb Protractors	180
Local Attraction	56
Lyons' Tables	202
Marking Pins	9, 144
Magnifying power of Telescope	40, 41
Magnets	188
Marine Glasses	193–195
Micrometer Telescope	95–97
Do or Stadia	115, 116
Microscopes	191–193
Miscellaneous	213

TABLE OF CONTENTS.

	PAGE.
Needle Instruments	56, 57
New York Rod	135
Object Glasses	215
Odometer	191
Opera Glasses	193–196
Optical principles of the Telescope	37–40
Outkeeper	25
Ovals	179
Palettes	203
Pantographs	185
Parallax, Instrumental	51
Parabolas	179
Parallel Rules	184
Pedometer	191
Pens	210, 211
Pencils	211
Philadelphia Rod	133
Pinion to Eye Glass	113
Plumbbobs	190
Pocket Compasses	6, 137, 138, 190, 191
Pocket Rules	185–187
Protractors, Swiss, German Silver	164–166
Do Horn	166, 167
Do Brass	167
Do German Silver	167
Do Paper	167
Do Ivory	167, 168
Do Boxwood	169, 170
Profile Paper	201
Do Books	202, 203
Prisms	215
Price List	1–11
Rack and Pinion	121, 122
Rain Gauge	214
Repairs to Instruments	16, 17
Do Compasses	25–27
Do Vernier Transit	52, 53
Reading Glasses	191–193
Refractions, Table of	102
Right Angle Sights	43
Rules	185-187
Rubber	212
Scales, Paper	171, 172
Sector Scales, Ivory, &c	168
Section Liners	184
Sextant	191

TABLE OF CONTENTS.

	PAGE.
Shifting Plate or Head	65
Size of Plain Compasses	27, 28
Do Vernier Compasses	33
Do Vernier Transits	55
Do Single Vernier Railroad Compass	62
Do Double Do Do	62
Do Surveyors' Double Vernier Transit	68
Do Do Single Do Do	69
Sights on Telescope	43
Sizes of Engineers' Transits	117
Do Y Levels	131
Do Builders', or Dumpy Level	132
Slabs and Saucers	203
Solar Attachment to Transits	5, 102–112
Solar Compasses, price of	5
Solar Compass	70–102
Spirit Levels	188, 189
Stencil Plates	216, 217
Straight Edges, Steel, Rubber, and Wood	174–177
Steel Goods	172–177
Surveyors' Cross	191
Swiss Drawing Instruments	146–152
Squares, Steel and German Silver	175–177
Tacks	185
Tapes, Chesterman's Metallic	9, 10, 145
Do do do Without Box	10
Do do Steel	10, 145
Do do do Pocket	10
Do Paine's Patent Steel	11
Tangent Scale	21
Telescope, how composed	35
Telescopes	197, 198
Theodolite Axis	118
Tramel Points	189
Tracing Paper	200, 201
Transits, Vernier	6, 34–55
Do Surveyors' 1. Vernier	6, 69
Do Do 2. Do	6, 63–68
Do Engineers' 2. Do	6, 113–118
Do Watch Telescope	6
Do Theodolite Axis	6
Do Extras to	7
Do Patent Solar Attachments to	5, 102–112
Triangular Boxwood Scales	170, 171
Do German Silver Scales	171
Triangles, Steel and German Silver	177
Do Rubber	178
Do Wood	179

TABLE OF CONTENTS.

	PAGE.
Tripods	139, 140
T Squares	180
Vara chains	8, 9, 141, 144
Variation of Needle	30
Variation, to turn off	31, 66
Vernier, to read the	30, 31
Verticle Circle	42
Vertical Arc, with moveable Vernier	104, 105
Vellum	200
Watch Telescope	117
Water Glasses	203
Water Colors, Winsor & Newton's	204
Do Box	205
Weights of Plain Compasses	28
Do Vernier Compasses	33
Do Vernier Transits	55
Do Single Vernier Railroad Compass	62
Do Double do do	62
Do Surveyors' Double Vernier Transits	68
Do do Single do	69
Do Solar Compass	101
Do Engineers' Transits	117
Do Y Levels	131
Do Builders', or Dumpy Levels	132
Do Patent Solar Attachment	112